The Climate Change Counter Movement

Ruth E. McKie

The Climate Change Counter Movement

How the Fossil Fuel Industry Sought to Delay Climate Action

palgrave
macmillan

Ruth E. McKie
De Montfort University
Leicester, UK

ISBN 978-3-031-33591-4 ISBN 978-3-031-33592-1 (eBook)
https://doi.org/10.1007/978-3-031-33592-1

Cover illustration: sorbetto/gettyimages

This Palgrave Macmillan imprint is published by the registered company Springer Nature Switzerland AG
The registered company address is: Gewerbestrasse 11, 6330 Cham, Switzerland

For Cola, my angel.
And
Dobby, the Little Devil

Acknowledgments

I would like to thank the support of my family for supporting me in my initial Ph.D. journey to my academic journey and producing this work. To my Ph.D. supervisors Paul Stretesky and Mike Long. I would not have written this without your support and confidence building during my Ph.D. and since. I would also like to the thank the feedback from scholars in the Climate Social Science Network (CSSN). I would particularly like to thank Kristofer Ekberg and Dieter Plehwe for their ongoing encouragement and support for the book. I would also like to thank Riley Dunlap who has championed my work since completing my Ph.D. Thank you all for your support.

About This Book

This book provides an insight into the emergence of climate change countermovement operations across the world. From the 1960s it became clear to the fossil fuel industry and related corporations that they would have to act fast to mitigate the impacts of fossil fuel production and consumption. However, because climate change posed a significant threat to business as usual from the late 1980s through the following two decades the climate change countermovement emerged to protect their industry. Moreover, it became the role of the climate change countermovement to influence policy and public opinion across countries to obstruct and delay effective and swift global action on climate change.

Contents

About the Author

Dr. Ruth E. McKie is a sociologist and received her Ph.D. in January 2018. The dissertation titled Rebranding the Climate Change Counter Movement through a Criminological and Political Economy Lens, is the basis for this book along with subsequent work in the field. Ruth's other research interests include animal abuse studies and feminist animal studies.

Abbreviations

ACSC	Australian Climate Science Coalition
AEF	Australian Environment Foundation
AEI	American Enterprise Institute
ALEC	American Legislative Exchange Council
ANFAVEA	National Association of Automotive Vehicle Manufacturers
API	American Petroleum Institute
ASSC	The Advancement of Sound Science Coalition
BP	British Petroleum
CAPP	Canadian Association of Petroleum Producers
CCCM	Climate Change Counter Movement
CCS	Center for Civil Society
CEI	Competitive Enterprise Institute
CEPA	Canadian Energy Pipeline Association
CEPOS	Center for Political Studies
CHC	Cooler Heads Coalition
CO_2	Carbon Dioxide
COP	Conference of the Parties
CPS	Centre for Policy Studies
CSCA	Climate Science Coalition of America
CSCCC	Civil Society Coalition on Climate Change
CSE	Citizens for a Sound Economy
DSP	Dominant Social Paradigm
EEI	Edison Electric Institute
EIKE	European Institute for Climate and Energy
EPA	Environmental Protection Agency
EPRF	Energy Probe Research Foundation

ESEF	European Science and Environment Forum
FCPP	Frontier Centre for Public Policy
FEC	Federal Election Campaign
FEE	Foundation for Economic Education
FFNZ	Federated Farmers
FMC	Fondazione Magna Carta
FOS	Friends of Science
GCC	Global Climate Coalition
GCSCT	Global Climate Science Communication Team
GDP	Gross Domestic Product
GES	Greening Earth Society
GHG	Greenhouse Gas Emissions
GWPF	Global Warming Policy Foundation
ICE	Information Council of the Environment
ICSC	International Climate Science Coalition
IEA	Institute of Economic Affairs
IER	Institute for Energy Research
IPA	Institute for Public Affairs
IPCC	International Governmental Panel on Climate Change
IPN	International Policy Network
IPPA	Initiative for Public Policy Analysis
IWF	Independent Women's Forum
MEP	Member European Parliament
MLI	Macdonald Laurier Institute
MPS	Mont Pelerin Society
NAM	National Association of Manufacturers
NCA	National Coal Association
NCPA	National Center for Policy Analysis
NDC	Nationally Determined Contributions
NEP	New Environmental Paradigm
NRSP	Natural Resource Stewardship Project
NZCPR	New Zealand Center for Political Research
NZCSC	New Zealand Climate Science Coalition
PACs	Political Action Committees
PERC	Property and Environment Research
PR	Public Relations
RAS	Russian Academy of Sciences
RSM	Rotterdam School of Management
SEPP	Science and Environmental Policy Project
SPN	State Policy Network
SRI	Stanford Research Institute
TASSC	The Advancement of Sound Science Coalition
UFM	Francisco Marroquin University

UK	United Kingdom
UN	United Nations
UNFCCC	United Nations Framework on Climate Change
US	United States
USA	United States of America
WFA	Western Fuels Association
WLF	Washington Legal Foundation

List of Tables

CHAPTER 1

Introduction

1.1 Fossil Fuel Capitalism and Ecological Degradation

Since the late fifteenth century, capitalism has been the hegemonic global economic system. In its more recent form, it is underpinned by the unsustainable exploitation of natural resources and subsequent ecological degradation (Foster et al., 2010), accompanied by a dominant worldview that shapes how humans, particularly in the West, interact with nature. In its simplistic terms under this view, nature's worth has been based primarily on how it can be exploited and used to fulfill the needs of human populations in the Western industrialized world, where natural resources are of monetary value only when removed from nature (Schnaiberg, 1980).

Historically, this exploitation of the environment included pillaging and theft of land from naturally resource-rich territories since the seventeenth century, alongside the permanent settlement of non-indigenous subjects through colonization of Global South countries. The outcome was that this practice laid the foundations for the relentless process of accumulating raw materials extracted, exploited, and commodified beyond meeting the individual needs of a community and instead entangled in an endless pursuit of economic profit. Moreover, this industrialization resulted in the unequal contribution to and the subsequent rise in greenhouse gas (GHG) emissions. Added to this, it made the

R. E. McKie, *The Climate Change Counter Movement*,
https://doi.org/10.1007/978-3-031-33592-1_1

conditions where those countries most exploited, with little responsibility for increasing GHG emissions, also became the most vulnerable to the impacts of climate change, with less economic capacity to respond and adapt without intervention from the country's most responsible.

From one perspective, this economic system encompasses complex social relations between individuals or social groups. As Huber (2009) notes, "fossil fuel energy is a much more *historically specific* mode of energy" (p. 106). During the period of the *Anthropocene*,[1] this mode of energy became central to social life because it connected with these social relations. These social relations of energy production have (1) centralized specific divisions of labor which require for their success, the exploitation and domination by one group. As Fast (2014) notes, "the relationship between capital and labor is based on exploitation is to recognize that capitalism is premised on the ability of one class of citizens to extract a surplus from another class of citizens" (p. 33). (2) A series of structural relations exist that place significant power and economic capital in the hands of those owners of production. Thus, the fruits of this economy are concentrated in the hands of a minority group of individuals.

From the 1980s onwards, the political and economic system transformed. It was spearheaded by the Thatcher and Reagan era in the UK and USA, whose governments pursued public policies promoting neoliberalization of the global economy. Neoliberalism is an ideology that subsequently impacts public policy design through an economic development strategy. This strategy promoted policies to increase privatization, specifically government-owned assets and national and international market deregulation. Moreover, it sought to act as a facilitator for businesses to operate without or with little government interference. Thus, companies like the fossil fuel industry could work with little to no state interference.

How neoliberalism emerged, and functions worldwide differed (Plehwe, 2015). Therefore, investigating a phenomenon of such complexity and its relevance to this story requires simplification by defining three central premises. (1) Capital is primarily concentrated in the hands of the few, the industry owners. (2) Ideology emphasizes the role of the market that emerged and aligned government policies to

[1] The Anthropocene is a geological age understood by the way humans have interacted with nature (Lewis & Maslin, 2015).

decentralized and privatized industries. Therefore, proposed and implemented government policies related to climate change would align with industry interests. (3) The construction of state/government policies primarily concerned deregulation and reducing government intervention in business activities. Notably, regulation is required to reconcile the role of overexploitation and the harmful impacts of industry on the environment. Therefore, limited regulation or deregulation in any policy, such as market-based solutions favoring the legacy of economic liberalism, reinforces an economic system that will exploit resources, increase GHG, and fail to negate the consequences of climate change.

Central to the success of this increasing wealth gap reinforcing class structures, and neoliberalization of the economy was the "hyper-materialistic" (Pérez & Esposito, 2010) consumer behavior of modern-day life. Through patterns of consumption that became inextricably linked to the identity of most groups in society (Wright & Nyberg, 2015), neoliberal capitalism served to provide a symbol of truth through which individual behavior aligned. In turn, it reinforced the notion that there is no other world without neoliberal capitalism, even despite what would become apparent was the unsustainable exploitation of resources and ecological decline. Given that the public must also respond to climate changes and support relevant policies, any response that challenged the individual behaviors intrinsically linked to the political economy would have to mediate these emotional and ideological connections. Thus, when business as usual and/or status quo is challenged, a response to protect it from those that most benefit not only has to target the institutions and structures that allow this political-economic structure to function, but it must also target the public.

1.2 Climate Change: Challenges to the "Status Quo"

Climate and earth scientists had documented changes caused by human actions to the Earth's atmospheric temperature since the early eighteenth century (e.g., Arrhenius, 1896; Callendar, 1938). However, attention on a global stage fully arrived in 1988 when the World Meteorological Organization and the United Nations Environment Program established the International Governmental Panel on Climate Change (IPCC). Shortly after the IPCC was established, the first international climate change conference took place in Rio de Janeiro at the United Nations Conference on Environment and Development (UN, 1992). At this conference, the first international treaty specifically addressed climate change, including Agenda 21 and the formation of the United Nations Framework on Climate Change (UNFCCC). It also created a global governance strategy designed exclusively to address the challenges faced by climate change. Since then, the UNFCCC has hosted yearly meetings where delegates from nation-states attend a series of negotiations, respond to scientific literature, establish a framework of targets, and implement a series of commitments based on a consensus—in its loosest sense—between countries. These international commitments were intended to play out on a national level where domestic-based, environmental, and transport-related policy decisions, along with other policy areas, go towards—in some way—meeting individual country commitments for reducing emissions and transitioning away from fossil fuels as the main source of energy. In turn and what was expected, was a global level effort to reduce GHG emissions and mitigate the impacts of climate change.

Over time, the UNFCCC evolved. Its substantial early focus was on climate mitigation for industrialized countries as part of the ratification of the 1997 Kyoto Protocol. On the surface, Kyoto was a landmark step for developing a collective global response to the climate crisis. However, it was not without its challenges, including the 2001 US withdrawal and overall failures to reduce global emissions (IPCC, 2014). Moreover, countries including China, India, and Brazil were excluded from the list. These countries have continued to increase GHG emissions through industrialization and development at a rapid rate since the 1980s. However, unlike other high-emitting countries, they were exempt from implementing further GHG reduction mechanisms. As explained in Chapter 6, in a historical context, these countries still bore less responsibility than those heavily polluting countries, such as the US, for their historical implications on human-caused climate change.

Similarly, in 2009 the negotiations at Copenhagen (COP15) failed to establish a "fair, ambitious and binding" agreement that would have laid the foundations for future coordinated action (Winkler & Beaumont, 2010, p. 639). The 2015 Paris Agreement was the next landmark for the UNFCCC, which sort to integrate established industrialized countries and those seeking to industrialize. This agreement included setting reductions in emissions targets across countries through nationally determined commitments (NDC). It also required increased financial support for loss and damage along with adaptation mechanisms, particularly for climate vulnerable countries in the Global South.

While governments, media, and international organizations claimed each yearly UNFCCC meeting was a success, the global response was fraught with tensions, inconsistencies, and obstruction. Obstruction here refers to a combination of ideological, economic, or political reasons and structures, institutions, and organizational actors that operated to impede a successful global response to climate change (adapted from Ekberg et al., 2022). The concept of obstruction can be separated into two levels. First, it can refer to a micro level, connected to how ideologies, structures, and institutions could affect individual behavior and subsequent individual responses to climate change or acts contributing to policy development and support, such as voting behavior. Second, at a macro-structural level, the concept refers to structural non-changes or changes that fail to challenge the irreconcilable relationship between economic growth defined under a measure of GDP and equally greater development in a neoliberal capitalist political economy.

In the context of this book, the fossil fuel industry and related corporate interests acted as the primary obstructive force at both global and national policymaking levels. Indeed, developing policies in this arena involved several stakeholders, including businesses, that engaged in negotiations and lobbying before such exchanges took place. While these actions take place practically—i.e., through negotiations on an international platform or lobbying to national delegates, which are then reflected in negotiations—disseminating information to influence public support for policies acts as another context in which obstruction manifests.

Garnering the public's attention and support for policies that sought to undermine climate action was arguably underpinned by the rise and acceleration of old and new environmental movements where climate change and climate justice became a key focus. This was especially relevant in the case of North America and Western Europe, whereby increasing knowledge of the impact of fossil fuels on the climate and responsibility lay particularly with these nations. This is because these new environmental movements and critical reflections on the implications of development and manifestation of the neoliberal capitalist political economy challenged the Dominant Social Paradigm (DSP) (Pirages & Ehrlich, 1974).

The DSP refers to an American-specific worldview shaped by the commitment to individualism and laisse-faire government, material abundance, the goodness of growth, faith in the efficacy of science and technology, and a view of nature as something to be subdued (Van Liere & Dunlap, 1981). This paradigm ultimately underpins the conditions in which fossil capitalism emerged and is maintained, whereby individualism and the need for material abundance require the unsustainable exploitation of nature. While initially traced to the USA, the DSP shapes the same worldviews that occupied most industrialized countries with high levels of human development (i.e., North America, Australia, and Western Europe) at a structural and individual level (Bogert et al., 2022). At a structural level, this reflects a society predicated on generating economic value through commodifying natural resources. Such a view then sheds light on the implications of overexploitation on the natural environment. At the individual level, it reflects individual consumer behavior. As Bogert et al. (2022) note, there are various degrees to which human–nature relationships are understood, and the DSP is situated among these. However, it is inherently affected by cultural and behavioral differences.

Recognizing these changes, Van Liere and Dunlap looked to develop a scale to understand these changes and devised the New Environmental Paradigm (NEP). This paradigm explored an alternate worldview shaped by ecological concern and, when evidenced by the rise of new environmental and social movements, oversaw the acceleration of a broader campaign to challenge the DSP. The DSP had previously organized behaviors around the hegemonic Western paradigm of fossil capitalism. Thus, in response to the clear evidence of environmental degradation caused by fossil fuel production and consumption, the public and, in turn, some policymakers began to reflect on the environmental harm caused by existing policies and the economy's structure.

This shift challenged the fossil fuel industry and how humans, particularly in the Global North, understood their relationship with nature. But it was not simply a relationship between *all* humans and nature. As Cotgrove (1982) contended, the DSP became dominant because it is based on views and ideas that align with dominant groups. First, the DSP, and in this case, the population predominantly in industrialized nations in the Global North, sees its relationship with nature based on the needs of the dominant group, legitimizing existing structures and institutions. In essence, it becomes an ideology that reinforces the status quo of fossil fuel use while reinforcing structural relations within a capitalist economic system. In other words, the DSP was entangled in a series of social relations of production whereby wealth is ultimately concentrated in the hands of an elite few. These include corporate leaders and the fossil fuel industry worth billions of dollars. Moreover, climate change became framed as primarily an issue of economics and development. And simultaneously, the environmental and climate justice movement's collective mobilization in response to structural injustices reflected the embedded unequal social and ecological relations under what is now known as a neoliberal [capitalist] economic system (Featherstone, 2015). Therefore, not only did human-caused climate change become a threat to business and profits, but it also posed a threat to the existing social order. As such, the shift away from the DSP meant that if the industry that underpins the world economy as we know it had to change, how do they then protect it?

1.3 Challenges to the Status Quo: The Emergence of a Climate Change Countermovement

As climate change became a central issue to local, national, and international governance, and public opinion showed a demonstratable appetite for change evidenced by the rise of new environmental movements, the fossil fuel industry responded by adopting a financial, political, and public relations (PR) strategy that bore similarities and even overlaps with the Tobacco Industry lobby (Oreskes & Conway, 2011). First, fossil fuel companies in and of themselves directly played a role by adapting their messaging strategy to undermine the evidence of climate change and the importance of the industry (Supran & Oreskes, 2021). Second, behind the scenes was the creation of a network of actors recruited to protect the interests of fossil capital. This broader network is the focus of this book: the Climate Change Counter Movement (herein CCCM). The CCCM can best be described holistically, as a multitude of organizations and individuals that, whether based on funding from fossil fuel interests, ideologically aligned, or promoted oppositional positions or deviations from the conventions of climate science, would forge a diverse collective effort to obstruct policy efforts and undermine climate science.

As Meyer and Staggenborg (1996, p. 1632) note, countermovements often adopted a similar formation to social movements, sharing many of the same "objects of concern" but making competing claims and vying "*for attention from the mass media and the broader public.*" Additionally, as Meyer (1993) notes, policy changes can alert to action and activate a largely disengaged public by elites. This means that elite groups can alert the public to political policy proposals or actions they believe the public will support—including those potentially disengaged on environmental issues. However, these may be critically based on this elite group's interests and not necessarily the public. Covered in this book, we see key moments where these interested groups as part of the CCCM, would exploit significant events in discourse that dominated their outputs and transcended into the media and political sphere. When social movements effectively create or exploit events, they are also likely to encourage countermovement mobilization advancing the causes of such elite groups. In effect, the CCCM organizations emerged when related policy activity questioned existing structures built on fossil capitalism. Thus, in this recollection, the CCCM was a strategically formed network initiative fostered in the fossil fuel industry's interests. Its primary aim was to cause conflict and devise and engage in activities to challenge environmental and climate justice movements, related climate legislation, and the transition away from fossil fuels, influencing politics and public opinion. A reactive network, it made contrary claims to social movements (i.e., environmental/climate movements) and formulated a collective consensus to resist the necessary social change to address the climate crisis.

This network of actors within the CCCM recounted in this book includes conservative, libertarian, and neoliberal think tanks, advocacy organizations or front groups, Conservative Foundations, and donor organizations, Coalition groups and organizations, trade associations, research institutes, economists, and related firms, PR firms, climate contrarians or global warming experts, media organizations and particularly the conservative media echo chamber.

Table 1.1 outlines definitions for each of the actors with the CCCM. While diverse in nature and definition across countries, it is important to note that some characteristics and relationships between each type of actor remain similar. This, in turn, shaped and forged what I describe as the global movement. First, organizational actors played a vital role in shaping policy, lobbying in the political realm, and disseminating information to the public and politicians on the issue of climate change. Think tanks played a particular role in legitimizing policy positions, some of which hold close ties to or have access to politicians and governments and those that fossil fuel companies may have directly funded. Helping to support and elevate opposition, advocacy groups would rally the public and lead on PR campaigns to spread the oppositional positions to the public. So, it was philanthropic foundations, explored more in Chapter 2, that allowed the operations of think tanks and advocacy organizations to flourish. Acting as a vehicle for funding and igniting campaigns, they played a vital role in maximizing outcomes for countermovement opposition. Directly relating to the fossil fuel industry are coalition organizations and trade associations that serve to represent their corporate interests. Some coalitions described in the book consisted of fossil fuel, transport, and energy-related industries. At the same time, trade associations would represent these broader industries and the individuals/employees that worked within them.

While organizational actors played an integral role in the CCCM, connecting them directly or indirectly are some specific individuals identified in this text can be labeled "climate contrarians." The idea of a climate contrarian is diverse across the literature. Still, as Petersen et al. (2020, p. 1) define, a contrarian is "a person that takes up a contrary position that is opposed to that of the majority." Like Petersen et al.'s methodological approach to defining contrarians, the individuals in this text who come to be defined as contrarians are drawn from the Heartland Institute's list of Global Warming Experts or those who have been Heartland Institute Climate Conference speakers (see Chapter 2). While some individuals would be defined as climate contrarians, other individuals would also take the form of TV personalities or disseminate ideologically aligned discourses of delay (discussed below). These individuals not only supported the PR campaigns of different organizations across the globe, but they also reproduced and translated works by organizations in different countries. In essence, these individuals acted as vehicles to diffuse the ideas of countermovement opposition between organizations and across countries.

Moreover, it becomes clear that in the case of this global countermovement, we can trace its roots to the USA. This is not a surprise, given as discussed in Chapter 2, the US arm of the countermovement in the shape and formation of today sparked and diffused its tactics across the world. With that said, the tactics of organizations and individuals and how this played out in different countries are also unique. Therefore, there remain similarities but also differences unique to the political-economic conditions of each country and what mechanisms to influence politics and or messages would work effectively promoted by CCCM organizations. Thus, each of these different movement actors played a vital role

Table 1.1 Definitions of counter movement organizational and actor types

Type of organization	*Definition*	*Example(s)*
Think Tanks	Think Tanks are political organizations that advise and develop comprehensive policy documents for governments worldwide (Fraussen & Halpin, 2017). Conservative think tanks, particularly those that have played a central role in climate obstruction movements, adopt a more advocacy-oriented position in immediate policy debates than their liberal counterparts (Weidenbaum, 2017). Conservative think tanks also have more financial support to promote their ideological interests over liberal and progressive think tanks (McGann, 2007). Consequently, these think tanks address policy issues only when they come to the forefront of political debates	Heartland Institute, Institute of Economic Affairs
Advocacy Organizations	Set up to undertake extensive lobbying activities and campaigns to recruit public support, advocacy organizations, or front groups can protect the interests of transnational and multinational corporations such as fossil fuel corporations (Dorsey, 2007)	Galileo Movement
Philanthropic Foundations	Philanthropic foundations act as third-party CCCM organizations distributing research grants or financial support to CTTs—the recipients of the most significant donations—and university-based research institutes. These helped mobilize CCCM campaigns by acquiring enough funding from charitable partners and donors (Brulle, 2014)	Koch Family Foundation, Sarah Scaife Foundation
Coalitions	In its simplest form, a coalition is when multiple individuals or organizations tend to have the same goals, research agenda, and political ideology (Axelrod, 1970). These coalition organizations are more likely to form when there is an opportunity to do so (Staggenborg, 1986). This means they are more likely to emerge under conditions of exceptional opportunity or threat, such as the rise of environmentalism and climate action	Cooler Heads Coalition

(continued)

Table 1.1 (continued)

Type of organization	*Definition*	*Example(s)*
Trade Associations	Trade association members donate or pay a subscription fee to the organization because the association aims to influence regulation and government policy for their interests (Rajwani et al., 2015)	Western Fuels Association
Climate Contrarians	Learning from the anti-environmentalism, wise use contrarians, these, in a sense, "celebrities" were used by the CCCM to act as authoritative voices in the scientific debate. These can also refer to what Oreskes and Conway (2011) called "protagonists [climate denial scientists] in the story of the CCCM that have merchandised doubt because they realized – with or without the help of academic decision theory – that doubt works. And it works in part because they have an erroneous view of science" (p. 267)	Fred Singer, Patrick Michaels, Ian Plimer
Economists and Economic consultancy Organizations	Economic consultancy firms provide analysis of current trends. In this case, these companies and individuals undertook specific forms of research, often although not always funded by industry to put forward particular analysis that would help and support the attempts of the fossil fuel industry to undermine action on climate change (e.g., Franta, 2022)	Charles Rivers Associates
Public Relations Organizations	Public relations firms conduct information and influence campaigns based on particular issues that can orient political decision-making. In the case of the CCCM, these PR campaigns can be performed by corporations and advocacy groups, or these same groups can employ PR companies specifically to engage in political activity and influence public opinion on the issue of climate change (Brulle & Werthman, 2021)	APCO World Wide
Conservative Media Echo Chamber	Elsasser and Dunlap (2013) describe the conservative media echo chamber as consisting of "major conservative TV, radio, and newspaper media (Brock, 2004; Jamieson & Cappella, 2008) but also a bevy of bloggers...and all other components of the denial machine in a mutual effort to undermine the reality and threat of global warming (Dunlap & McCright, 2011)" (p. 755)	Fox News, Rush Limbaugh, Andrew Bolt

in galvanizing the campaign to undermine climate action working alongside and on behalf of the fossil industry to challenge international and national legislation and transform and shape public opinion. To ultimately gain success, it was down to the CCCM to support the fossil industry, constructing a narrative to counter environmental movements that aimed in some way to change society's relationship with nature.

Central to the success of the countermovement was the requirement to construct an effective messaging strategy that could be reproduced or redeveloped to fit a particular audience. The CCCM would be able to target arguments that reflected each location's political-economic contexts, whether through industry-specific arguments, reflecting on developmental needs, to ideologically framed positions that spoke to both the public and politicians who would need to support progressive climate-related policy. Evidence throughout, these movement actors also utilized political opportunities (Meyer & Minkoff, 2004), promoting and elevating arguments at particular points to coincide with key climate-related events such as political controversies or international conferences. While there is variation between arguments that would be used in this global movement, researchers have identified several arguments to describe the discourse emerging from the CCCM overall (e.g., Dunlap & Jacques, 2013; Jacquet et al., 2014; Jylhä et al., 2016; Lewandowsky et al., 2013; McKie, 2018, 2019, 2022; Norgaard, 2011; Stoll-Kleemann et al., 2001). In its most straightforward form, arguments proposed by the CCCM can be divided into two categories. One includes science-based arguments, which focus on questioning the scientific evidence on climate change, whether that be denying climate change exists or proposing there are benefits to rising CO_2 emissions. Two, strategic or policy-based arguments that look at the creation and implications of climate change policy.

Ultimately, the spectrum of arguments constructed by the movement was multifaceted and Lamb et al. (2020) identified four broad categories of "*discourses of delay*" that are relevant here. The first category *redirecting responsibility* refers to the assumption that someone else should take responsibility. This could be "individuals" rather than fossil fuel corporations changing their behavior or comparing actions of one country to those of another, whereby any action taken in isolation would disadvantage the nation and/or have limited to no impact. The second category *Pushing non-transformative solutions* refers to an assumption that technology will unlock multiple possibilities to address climate change—a perception often associated with a model of ecological modernization. This technological optimism also included fossil fuel solutionism, where continued fossil fuel exploitation would help address the climate crisis.

Moreover, these non-solutions should be voluntary and reflect a form of free market environmentalism. Adopting this ideological approach by relying on market forces to address climate change would thus maintain business as usual under a neoliberal political economy. The third category *emphasizes the downsides of action*, presenting a case that (1) continued use of fossil fuels is essential for both the development and wellbeing of the global population, (2) continued fossil fuel use is necessary to help vulnerable countries where climate-related policies will inflict harm on the population, and (3) these lead to the idea of a need for "policy perfectionism." Finally, the fourth category centers on *surrendering* because climate change is impossible to mitigate. Because attempting to mitigate the crisis is the antithesis to dominant ways of life, climate change is locked in, and climate changes have deemed the fate of God.

As explained in this book, actors within the CCCM deployed these discourses of delay. CCCM actors used these discourses to frame policy proposals that could be integrated into policy at local, national, and international levels, seep into public perceptions affecting individual behavior and the ballot box, and electing non-progressive parties or individuals with little concern for climate change.

1.4 Structure of the Book

Before moving further, it is crucial to clarify the approach to understand how all organizations and actors can be understood in the context of a global countermovement. As such, there are three key points to note. First, each chapter is structured to illuminate a network of actors within a country. Chapters identify interlocks between individuals and organizations that played a role in facilitating or engaging with activities that challenged the climate consensus using examples. In turn, it illustrates the complexity of how the CCCM operated within each location. Second, each chapter assessment reflects what I argue is a *pluriverse* of CCCM organizations. Escobar (2015) describes the pluriverse as "a space for the study and advancement of transitions toward a world in which many worlds are in it" (p. 14). It refers to the process of *Transiciones*, representing a global spread of the idea that one single model can explain a phenomenon. Therefore, understanding the emergence of

CCCM across different geographic locations is complex, yet there remains some overlapping consistency.

Third, my work so far only serves as an introduction to help us understand countermovement opposition over time in specific geographic locations. It draws on my own archival research and the work of other scholars, investigative journalists, non-governmental organizations, and activists that have contributed significantly to our understanding of the CCCM in different countries. It is by no means an exhaustive account of countermovement operations. After reading this text, I urge the reader to continue exploring the topic, examining the work of those featured and ongoing research in the field.

The structure of the book is as follows. I examine different geographic locations from the late 1990s–2016 to trace the countermovement's changing dynamics, identifying core, important organizations, and unique examples from eight years of academic research. Ultimately, several elements of the CCCM and climate obstruction more broadly are not explored here, and this work complements others. Moreover, I present the discussion with the intention of the past and a countermovement that no longer exists. However, at the time of writing, many of these organizations continue to operate and impact politics and public opinion.

I begin by turning to the countermovement's roots in the United States of America (USA) (Chapter 2). This chapter recounts the formation of a small network of organizations predominantly of the fossil fuel industry and their trade associations that would later help facilitate a more extensive scale network through foundation funding and media outreach. The purpose of this network was to impact public policy, create and foster skepticism within the public mind, and protect the "status quo." However, the "ideas" that could not stall effective action across countries would be ineffective without the influence of European thinkers. As a result, Chapter 3 turns to the operations of think tanks that would disseminate discourses of delay into the political and public sphere, continuing the movement's expansion to prevent a global shift. Here, I demonstrate clear overlaps and the formation of transnational relationships that link with the actions of CCCM in Australia and New Zealand (Chapter 4) and Canada (Chapter 5). Chapters 4 and 5 examine the specific social, cultural, and economic identities related to corporate sectors of coal, gas, and agribusiness that shaped some of the actions of the CCCM. Finally, in Chapter 6, I attempt to address gaps in our knowledge of the history of the CCCM in the Global South. The chapter highlights some early examples of countermovement opposition in Global South countries. However, this is only the beginning of understanding CCCM resistance in this country. We still have much to learn about the complexities and contours of countermovement activity.

References

Arrhenius, S. (1896). XXXI. On the influence of Carbonic Acid in the Air upon the temperature of the ground. *The London, Edinburgh, and Dublin Philosophical Magazine and Journal of Science, 41*(251), 237–276.

Axelrod, R. M. (1970). *Conflict of interest: A theory of divergent goals with applications to politics*. Markham Publishing Company.

Bogert, J., Ellers, J., Lewandowsky, S., Balgopal, M., & Harvey, J. (2022). Reviewing the relationship between neoliberal societies and nature: Implications of the industrialized dominant social paradigm for a sustainable future. *Ecology and Society, 27*(2), 7.

Brulle, R. J. (2014). Institutionalizing delay: Foundation funding and the creation of US climate change countermovement organizations. *Climatic Change, 122*(4), 681–694.

Brulle, R. J., & Werthman, C. (2021). The role of public relations firms in climate change politics. *Climatic Change, 169*(1), 1–21.

Callendar, G. S. (1938). The artificial production of carbon dioxide and its influence on temperature. *Quarterly Journal of the Royal Meteorological Society, 64*(275), 223–240.

Cotgrove. (1982). *Catastrophe or Cornucopia*. Wiley.

Dorsey, M. K. (2007). Climate knowledge and power: Tales of skeptic tanks, weather gods, and sagas for climate (in) justice. *Capitalism Nature Socialism, 18*(2), 7–21.

Dunlap, R. E., & Jacques, P. J. (2013). Climate change denial books and conservative think tanks: Exploring the connection. *American Behavioral Scientist, 57*(6), 699–731.

Ekberg, K., Forchtner, B., Hultman, M., & Jylhä, K. M. (2022). *Climate obstruction: How denial, delay and inaction are heating the planet*. Taylor & Francis.

Elsasser, S. W., & Dunlap, R. E. (2013). Leading voices in the denier choir: Conservative columnists' dismissal of global warming and denigration of climate science. *American Behavioral Scientist, 57*(6), 754–776.

Escobar, A. (2015). Transiciones: A space for research and design for transitions to the pluriverse. *Design Philosophy Papers, 13*(1), 13–23.

Fast, T. (2014). Stapled to the front door: Neoliberal extractivism in Canada. *Studies in Political Economy, 94*(1), 31–60.

Featherstone, D. (2015). The contested politics of climate change and the crisis of neo-liberalism. *ACME: An International Journal for Critical Geographies, 12*(1), 44–64. https://acme-journal.org/index.php/acme/article/view/951. Accessed 24 Oct 2022.

Foster, J. B., Clark, B., & York, R. (2010). *The ecological rift: Capitalisms war on the earth.* Monthly Review Press.

Franta, B. (2022). Weaponizing economics: Big Oil, economic consultants, and climate policy delay. *Environmental Politics, 31*(4), 555–575. https://doi.org/10.1080/09644016.2021.1947636

Fraussen, B., & Halpin, D. (2017). Think tanks and strategic policy-making: The contribution of think tanks to policy advisory systems. *Policy Sciences, 50*(1), 105–124.

Huber, M. T. (2009). Energizing historical materialism: Fossil fuels, space and the capitalist mode of production. *Geoforum, 40*(1), 105–115. https://doi.org/10.1016/j.geoforum.2008.08.004

International Panel on Climate Change. (2014). *Summary for policy makers.* https://www.ipcc.ch/report/ar5/wg2/summary-for-policymakers/. Accessed 24 Oct 2022.

Jacquet, J., Dietrich, M., & Jost, J. T. (2014). The ideological divide and climate change opinion: "Top-down" and "bottom-up" approaches. *Frontiers in Psychology, 5*, 1458.

Jylhä, K. M., Cantal, C., Akrami, N., & Milfont, T. L. (2016). Denial of anthropogenic climate change: Social dominance orientation helps explain the conservative male effect in Brazil and Sweden. *Personality and Individual Differences, 98*, 184–187.

Lamb, W. F., Mattioli, G., Levi, S., Roberts, J. T., Capstick, S., Creutzig, F., Minx, J. C., Müller-Hansen, F., Culhane, T., & Steinberger, J. K. (2020). Discourses of climate delay. *Global Sustainability, 3*, e17.

Lewandowsky, S., Oberauer, K., & Gignac, G. E. (2013). NASA faked the moon landing—Therefore, (climate) science is a hoax: An anatomy of the motivated rejection of science. *Psychological Science, 24*(5), 622–633.

Lewis, S. L., & Maslin, M. A. (2015). Defining the Anthropocene. *Nature, 519*(7542), 171–180.

McGann, J. G. (Ed.). (2007). *Think tanks and policy advice in the US: Academics, advisors and advocates.* Routledge.

McKie, R. E. (2018). *Rebranding the climate change counter movement through a criminological and political economic lens* [Doctoral dissertation, Northumbria University]. https://nrl.northumbria.ac.uk/id/eprint/33466/. Accessed 24 Oct 2022.

McKie, R. E. (2019). Climate change counter movement neutralization techniques: A typology to examine the climate change counter movement. *Sociological Inquiry, 89*(2), 288–316.

McKie, R. E. (2022). Climate change counter movement organizations: An international deviant network? In *Handbook of anti-environmentalism* (pp. 173–191). Edward Elgar Publishing.

Meyer, D. S. (1993, June). Institutionalizing dissent: The United States structure of political opportunity and the end of the nuclear freeze movement. In *Sociological Forum* (Vol. 8, No. 2, pp. 157–179). Kluwer Academic Publishers-Plenum Publishers.

Meyer, D. S., & Minkoff, D. C. (2004). Conceptualizing political opportunity. *Social Forces, 82*(4), 1457–1492.

Meyer, D. S., & Staggenborg, S. (1996). Movements, countermovements, and the structure of political opportunity. *American Journal of Sociology, 101*(6), 1628–1660.

Norgaard, K. M. (2011). *Living in denial: Climate change, emotions, and everyday life*. MIT Press.

Oreskes, N., & Conway, E. M. (2011). *Merchants of doubt: How a handful of scientists obscured the truth on issues from tobacco smoke to global warming*. Bloomsbury Publishing.

Pérez, F., & Esposito, L. (2010). The global addiction and human rights: Insatiable consumerism, neoliberalism, and harm reduction. *Perspectives on Global Development and Technology, 9*(1–2), 84–100.

Petersen, A. M., Vincent, E. M., & Westerling, A. L. (2020). Addendum: Discrepancy in scientific authority and media visibility of climate change scientists and contrarians. *Nature communications, 11*(1), 553.

Pirages, D. C., & Ehrlich, P. R. (1974). Ark II; social response to environmental imperatives [by] Dennis C. Pirages [and] Paul R. Ehrlich.

Plehwe, D. (2015). Introduction. In P. Mirowski & D. Plehwe (Eds.), *The Road from Mont Pèlerin: The making of the neoliberal thought collective, with a new preface* (pp. 10–42). Harvard University Press.

Rajwani, T., Lawton, T., & Phillips, N. (2015). The "Voice of Industry": Why management researchers should pay more attention to trade associations. *Strategic Organization, 13*(3), 224–232.

Schnaiberg, A. (1980). *The environment: From surplus to scarcity*. Oxford University Press.

Supran, G., & Oreskes, N. (2021). Rhetoric and frame analysis of ExxonMobil's climate change communications. *One Earth, 4*(5), 696–719.

Staggenborg, S. (1986). Coalition work in the pro-choice movement: Organizational and environmental opportunities and obstacles. *Social Problems, 33*(5), 374–390.

Stoll-Kleemann, S., O'Riordan, T., & Jaeger, C. C. (2001). The psychology of denial concerning climate mitigation measures: Evidence from Swiss focus groups. *Global Environmental Change, 11*(2), 107–117.

United Nations. (1992). United Nations Conference on Environment & Development Rio de Janeiro, Brazil, 3 to 14 June 1992. https://sustainabledevelopment.un.org/content/documents/Agenda21.pdf. Accessed 22 Nov 2022.

Van Liere, K. D., & Dunlap, R. E. (1981). Environmental concern: Does it make a difference how it's measured? *Environment and Behavior, 13*(6), 651–676.

Weidenbaum, M. (2017). *The competition of ideas: The world of the Washington think tanks*. Routledge.

Winkler, H., & Beaumont, J. (2010). Fair and effective multilateralism in the post-Copenhagen climate negotiations. *Climate Policy, 10*(6), 638–654.

Wright, C., & Nyberg, D. (2015). *Climate change, capitalism, and corporations*. Cambridge University Press.

CHAPTER 2

The Foundations of the Climate Change Counter Movement: United States of America

2.1 Introduction

Beginning in the late 1980s, the fossil fuel industry accelerated a campaign to undermine climate action by governments and the public. This started a multi-decadal campaign through well-funded networks to protect an industry under threat. This chapter describes the formation of the CCCM, first looking at the early and organized opposition from coalition groups of fossil fuel companies and other interested parties to the emergence of a network of think tanks predominantly funded by fossil fuel corporations and conservative donors. It then explores how these organizations successfully helped create doubt that led to inaction during the George H.W. Bush administration (2001–2008) before escalating their campaign throughout the Barack Obama administration (2008–2016). This is because Obama attempted to make marked progress on climate change. Still, these would be contested and undermined, leading to the USA's evident failure to take progressive steps to address the climate crisis.

2.2 The Fossil Industry Activated (1950–1980)

In the late 1950s, the fossil fuel industry began its coordinated effort to undermine the environmental and later climate-related legislation to protect the industry from increasing regulation, oversight, and accountability. First, fossil fuel companies, including Exxon, previously Humble

R. E. McKie, *The Climate Change Counter Movement*,
https://doi.org/10.1007/978-3-031-33592-1_2

Oil, would commission and conduct independent studies which revealed the link between the combustion of fossil fuels and CO_2 emissions (Center for International Environmental Law, 2016). Subsequently, these corporations would commission and fund research, outside organizations, and individuals endeavoring to craft a network of actors to undermine climate legislation and influence public opinion. Thus, starting the campaign to undermine climate science, climate legislation, and the development of the CCCM.

The American Petroleum Institute (API) was one of the first key players in developing the CCCM. It was the *"primary trade association for the oil and gas industry in the USA"* (Dickson, 2007, p. 2), and, over time, accrued around 400 corporate members, including Exxon Mobil, Chevron, and British Petroleum (BP). The API actively lobbied, produced, and commissioned research, engaged in PR activities to support the fossil fuel industry's interests, and generated industry policy. Moreover, it shared a close relationship with government agencies and interest groups that sought to influence environmental legislation and regulation. For example, in 1968, the institute commissioned Stanford Research Institute (SRI) scientists Elmer Robinson and R.C. Robbins to produce a report on the impact of fossil fuel use and CO_2 emissions. The authors summarized that CO_2 emissions from fossil fuels were outstripping that of natural CO_2. Between the 1970s and early 80s, several more reports and briefings documenting the risks from CO_2 emissions were distributed among fossil interest groups. Subsequently, between 1979 and 1983, a task force was set up made up of fossil fuel companies, including Exxon, Mobil, Amoco, Phillips, Texaco, Shell, Sunoco, Sohio and Standard Oil of California and Gulf Oil (two companies that became Chevron) facilitated by the API to discuss the science and implications of climate change on the oil industry (Bannerjee, 2015).

By identifying interested individuals and organizations that wanted to prevent increasing regulation on the oil, gas, and energy sectors, ExxonMobil and Koch Industries, along with others, would fund a network of organizations to engage in lobbying efforts and a PR campaign to reach the public. This network included conservative and neoliberal think tanks; advocacy organizations or front groups; conservative foundations and donor organizations; coalition organizations; trade associations; research institutes; professional associations; economists and related companies; public relations firms; scientists; politicians; the Republican party; and the conservative media echo chamber. Ultimately, this formulation of organized activities by different types of organizations forged what we understand as the CCCM in its broadest sense.

Fundamental to these operation networks were financial contributions. Fossil fuel corporations, conservative and family foundations, and other interested parties donated to "non-political" organizations such as think tanks that became a primary voice of the CCCM in the USA. These donors included but were not limited to Dunn's Foundation for Right Thinking, Sarah Scaife Foundation, Lynne Foundation, and Koch Foundations (e.g., Brulle et al., 2021). While it is impossible to make explicit connections between the funding from these foundations and climate-related content, we can see political positions, individuals, and financial overlaps between foundations and other countermovement actors in a broader context.

Uncovering funding network sources was made more difficult following amendments to funding declarations within the US political system under Citizens United in 2010. Citizens United v. Federal Election Commission was a landmark reform in campaign financing restrictions. Subsequently, wealthy donors and corporations (including fossil fuels companies) could fund political campaigns without identification under the premise of "free speech," gaining increased influence over the US political system (see Levitt, 2010 for further information on Citizens United). These amendment changes meant that organizations could retain anonymity and spend unlimited funds supporting a political candidate or party that could influence legislation to protect their industries. This process was better known as *dark money*. In addition, Donors' Trust and Donors' Capital were created to distribute funding from individuals and foundations. The foundations made grants based on donor preferences while remaining anonymous.

Despite this secrecy, Brulle (2014) and Brulle et al. (2021) summarized and explored these financial ties, including conservative foundations and Donors' Trust and Capital. Between 2003 and 2018, donations data from the identifiable foundations to 128 CCCM organizations saw funding increased from $357 million to $808 million, peaking at $811 million in 2012. Of the named donors, some included the Scaife Family Foundation ($125 million) and Charles Koch Institutes and Foundations ($74 million). This provided CCCM organizations with the financial backing to operate. The critical question is, what drove the connections and organizations to work within this network?

The CCCM used some of the same blueprints of the anti-environmental movement of the 80s in line with the advancement of neoliberalism and economic changes under Reagan (Antonio & Brulle, 2011). Central was building the argument that new and existing regulations ultimately stalled economic growth and changes. Therefore, any changes should not increase regulation but instead the opposite, expanding free enterprise and private property rights (Brick, 1995). Peeples' (2014) described the Wise Use Movement as a coalition of front groups that ultimately targeted regulation of federal lands. The movement promoted arguments that would aim to enhance private property rights and reduce government regulation of publicly held property so that they could be privatized and subsequently commodified either for personal or corporate gain. To do this, the arguments associated with private property rights and this approach to prevent regulation were perceived as some form of conservation and a new form of environmentalism. This argument previously invoked by Wise Use front groups and supporters was also central to the interests of fossil capitalism and intertwined particularly with right-wing (Republican) political organizations that became part of the CCCM. Thus, from the late 1980s onwards, without uncertainty, the fossil industry and related organizations acknowledged and understood the harm of CO_2 emissions from the fossil fuel industry on the climate.

2.3 Proliferation of the Counter Movement: 1990s–Early 2000s

The fossil fuel industry would recruit economists to produce reports and offer recommended "solutions" in response to (not) addressing climate change that would favor their interests (Franta, 2021). Franta (2021) identified how the fossil industry and a select number of economists between the 1990s–2010s produced biased economic analyses to undermine climate policies in the USA. One of the first examples was in 1991 when the API commissioned a report by US-based financial consulting firm Charles River Associates. Charles Rivers Associates received funding from the fossil fuel industry and produced outputs that would help: "convince the public and policymakers that climate policy would be costly, global warming would be relatively unimportant, and there would be little harm in delaying action" (p. 558). A 1991 report authored by David Montgomery, who had previously served in US administrations under

Presidents Ford, Carter, and George H.W. Bush, was disseminated in a media campaign to challenge the implementation of climate policies based on the premise of devastating economic impacts. The executive vice president at the time, William O'Keefe would inaccurately report the findings and refrain from clearly disclosing it had been at least partially funded by the fossil fuel industry.

O'Keefe, went on to serve as Chairman of the Global Climate Coalition (GCC), and continued to use the work of economists such as Montgomery to gain further support to challenge the implementation of policies that could increase regulation on the oil and gas sector. The GCC (1989–2002) was the first of many coalitions formed to represent the industry. It launched an active campaign to undermine climate policy often with overlapping directorship, donorship, and similar policy-related outputs and positions on climate change and regulation. Operating from within the National Association of Manufacturers (NAM) building—a trade association that produced reports and campaigns to undermine IPPC negotiations while promoting and elevating the voice of Patrick Michaels, Fred Singer and Richard Lindzen—the GCC became an independent non-profit organization in 1995. Member organizations included the API, Chevron, Exxon, and General Motors, to name a few. The GCC ignited a PR campaign and intense lobbying activities, even gaining direct access to political administrations. In 2001, for example, representatives participated in a meeting with the George W. Bush administration, which later withdrew the USA from Kyoto negotiations. An unclassified document released from the meeting revealed, "*Potus reject Kyoto, in part, based on input from you*" (Vidal, 2005, n.p.). Responding to the political opportunity of a changing presidency and the implementation of the Kyoto Protocol, these actors were able to create doubt that the progress in addressing climate change under this current international governance strategy would hinder US growth and prosperity.

After the demise of the GCC, several of the same interested parties formed other coalition groups with changing membership over time (Jones et al., 2019) (Table 2.1). For example, several GCC members formed the Alliance for Climate Strategies, including the Edison Electric Institute (EEI) trade association. Later some of these organizations became part of the coalition Americans for Balanced Energy Choices which focused on coal-fired electricity, receiving financial support from Peabody Energy, Duke Energy, and the National Rural Electric Cooperative Association. The Cooler Heads Coalition (CHC) (1997–present), created by the now-defunct *National Consumer Coalition* (1997–2004) and part-funded by *Consumer Alert Inc* (1977–2005) also became a leading voice in countermovement opposition. Consumer Alert Inc was an American non-profit primarily funded by corporations including Exxon Mobil with a specific amount of grant funding earmarked for climate change-related programs (Climate Investigations, n.d.).

The CHC comprised conservative, neoliberal, and libertarian think tanks—some of which were funded by fossil fuel corporations and conservative foundations—which helped elevate the voice of fossil interests into policy in the USA. While there were organizations outside of the USA that made up part of the coalition (See Table 2.2), the CHC engaged

Table 2.1 US Coalition Groups (adapted from Jones et al., 2019)

Coalition group name	*Years in operation*	*Important people and organizations*
Global Climate Coalition	1989–2002	William O. Keefe, EEI, WFA
Cooler Heads Coalition	1997–Present	Myron Ebell, Competitive Enterprise Institute, Koch Industries, Exxon Mobil
Americans for Balanced Energy Choices	2000–2008	
Center for Energy and Economic Development	1992–2008	Western fuels Association, Peabody Energy
American Coalition for Clean Coal Electricity	2008–present	Western Fuels Association, Kentucky Coal Association
Alliance for Energy and Economic Growth	2001–2011	API, EEI, Jay Timmons (National Association of Manufacturers)
Coalition for Affordable and Reliable Energy	2001–2008	EEI, Peabody Energy Paul Oakley
Information Council on the Environment	1991–	Robert Balling, Patrick Michaels, Sherwood Idso
Coalition for Vehicle Choice	1991–early 2000s	American Legislative Council, Competitive Enterprise Institute, Citizens for Sound Economy
Partnership for a Better Energy Future	2014–present	NAM, Global Energy Institute, Jay Timmons
Coalition for American Jobs	2010–2013	API, Marty Durbin (API executive vice president)
Alliance for Climate Strategies	2002–2007	Glenn Kelly, Edison Electric Institute

Table 2.2 Member organizations of the cooler heads coalition

Organization	*Country*
60 Plus Association	USA
Alexis de Tocqueville Institution	USA
Americans for Tax Reform (ATR)	USA
Americans for Prosperity (AFP)	USA
American Policy Center (APC)	USA
America's Future Foundation (AFF)	USA
Committee for a Constructive Tomorrow (CFACT)	USA
Competitive Enterprise Institute (CEI)	USA
Fraser Institute (Canada)	Canada
Frontiers of Freedom	USA
George C. Marshall Institute	USA
Heartland Institute	USA
Independent Institute	USA
Bruno Leoni Institute	Italy
Junkscience.com	USA
Lavoisier Group	Australia
Liberty Institute	India
National Center for Policy Analysis (NCPA)	USA
Pacific Research Institute (PRI)	USA
Seniors Coalition	USA
Small Business and Entrepreneurship Council (SBEC)	USA

in PR activities, sponsored economic and scientific research and briefings, led educational initiatives, and hosted events where it would invite members of Congress, policymakers, and the media. Consumer Alert created the organization to dispel "the myths of global warming by exposing flawed economic, scientific, and risk analysis...and follow the progress of the international Global Climate Change Treaty negotiations" (CHC, n.d.). One of its main contributors and leaders was Myron Ebell. Ebell the director of the Global Warming and International Environmental Policy sector at the CEI published articles and op-eds that could reach the mainstream media and led the organization. He also had other ties to other CCCM including the Heartland Institute and would make appearances on TV, films, and documentaries disseminating discourses of delay (IMDB, n.d.). Moreover, the CHC website was funded by the Competitive Enterprise Institute (CEI) (CHC, n.d.).

Accompanying coalition groups, the Chamber of Commerce, and the API, were other trade associations for US coal, oil, gas, electricity, rail, and automobile sectors. This included Western Fuels Association (WFA), a member of several advocacy groups and coalitions, including the Information Council of the Environment (ICE). The ICE was formed with, among others, the Southern Company and the EEI, to develop a national communication and PR campaign designed to undermine climate science and influence the administration's approach to climate and energy policy. As Jones et al. (2019) noted, "while little information about ICE exists beyond the leaked documents describing the market testing campaign, it is clear that ICE pioneered frameworks for mass denial activities that are still incredibly relevant today" (p. 9).

The WFA often challenged proposed climate policies or regulatory changes affecting the fossil fuel industry. The positions they promoted were created for lobbying activities and disseminated to its members to discredit and minimize the risks associated with climate change, campaigning to reduce its responsibility for the costs of coal burning and advocate against climate regulation. This was evidenced by the creation of the PR organization the Greening Earth Society (GES) to promote the view that fossil fuels enable economic activity and that CO_2 level rises are better for the Earth (Greening Earth Society, n.d.). Advisors to the GES were individuals that had a record of questioning either the evidence of human-caused climate change, undermine the impact of humans or critique and attacking policymakers and/or environmental activists and organizations. The organization also released the controversial videotape in 1992 titled *The Greening of Planet Earth,* diagnosing the benefits of increased concentrations of CO_2 on plant life, followed by a second *The Greening of Planet Earth Continued* in 1998.

To ensure success, the conservative echo chamber played a vital role in platforming organizations and individuals deploying discourses of a delay from these organizations to spread messages to the US public. Primarily, this echo chamber existed through traditional media, including newspapers, radio, and television, frequently with partisan affiliations to the political right. This was in no way a simple accident. Indeed, it started in the early stages of the CCCM. For example, in 1991, the Information Council for Environment (ICE), composed of trade associations including The National Coal Association (NCA), WFA, and EEI, created a media plan to reposition global warming as "theory"—disputing the scientific evidence—and throughout its history spent around $500,000 on an advertising campaign (Mulvey & Shulman, 2015). Activities included placing specific ads alongside Am Talk Radio, notably the conservative Rush Limbaugh Show. Then, in 1998, a memo was sent from the Global Climate Science Communication Team (GCSCT), a group convened by the API (Climate Files, n.d.) to fossil fuel industry executives about the proposed media strategy. Members of the GCSCT included: John Adams (John Adams Associates), Candace Crandall (Science and Environmental Policy project), David Rothbard (CFACT), Jeffery Salmon (The Marshall Institute), Lee Garringan (Environmental Issues Council), Lynn Bouchey (Frontiers for Freedom), Myron Ebrell (Frontiers for Freedom), Peter Cleary (Americans for Tax Reform), Randy Randol (Exxon Corp), Robert Gehri (The Southern Company), Sharon Keneiss (Chevron Corp), Steven Milloy (TASSC), Joseph Walker (API) (Inside Climate, n.d.). In this memo, the team stated they would recruit scientists for media outreach. This action aimed to legitimize the opposition positions adopted more broadly by different countermovement organizations, including producing fact sheets and help kits that could be distributed to the public.

Conservative commentators such as Rush Limbaugh had a significant presence in shaping Republican Party discourse and ideas to the public. His radio talk show, TV, and online news channel, was for 30 years a place to disseminate traditional climate skepticism, particularly to Conservative and Republican supporters. As Carmichael et al. (2017) note, when Limbaugh spent more time on the issue of climate change, Republicans were significantly less concerned. Moreover, they noted that the conservative media echo chamber would increase distrust in scientists and thus the certainty of climate change science.

While early evidence of conservative echo chambers emerged on the radio and in newspapers, things certainly changed with the expanding role of TV, notably The Rupert Murdoch-owned outlet News Corp's Fox News (see also Krosnick & MacInnis, 2010). Fox News became the core medium for disseminating discourses of delay to the American public which would benefit countermovement opposition. McKnight (2010) recalled the political influence of News Corp and Murdoch, outlining that News Corp was a vehicle for diffusing a political and cultural agenda, "supporting or opposing a particular party or decision" (p. 304) and advancing certain ideological beliefs. These beliefs centered on a free market ideal and economic determinism central to the political economy in the USA. Hence, News Corp outlets disseminated ideas to shape public worldviews that naturalize laisse-faire economics and reinforce the hegemony of fossil fuel capitalism and the DSP.

This is unsurprising given Murdoch was an MPS fellow (Chapter 6) and News Corp outlets were disseminating the same ideas. Both being advocates of neo-liberalization which underpins the anti-environmental ideological positions of the CCCM. Moreover, in 1997, Murdoch joined the board of the Cato Institute, an organization founded by Charles Koch (of Koch Industries), Murray Rothbard, and Ed Crane. Not only with existing ties to the Koch family, the Cato Institute also had clear markers of activity that place it within the CCCM. With a vision "to create free, open, and civil societies founded on libertarian principles" (Cato Institute, n.d.-a), Cato's libertarian ideological foundations stood in opposition to increased regulation, government intervention, and the collective action of environmental movements that swept across the USA from the 1960s. Indeed, the collective mobilization of environmental movements stood in stark contrast to the individualism underpinning the organization's ideology, which also played a role in the rise and influence of the Wise Use Movement in the 1980s (Brick, 1995). These arguments certainly mirrored and reflected what would later be utilized by CCCM organizations.

Moreover, the work of Cato scholars featured in Murdoch-owned newspapers and on Fox News (McKnight, 2010). Examples included a 2009 Cato Institute ad on a policy report that featured well-known individuals who vocalized oppositional positions or discourses of delay (Cato Institute, 2009a) and Patrick Michaels, Senior Fellow at Cato, was featured in 2009 on Fox News "News Room" (Cato Institute, 2009b). Using media ties with Murdoch organizations, Cato was critical in elevating the messages of the countermovement. Moreover, this operation created a sense of "pseudo-authority" (p. 264), whereby these individuals operating within and connected to CCCM organizations could present arguments in news outlets. And the Cato Institute would provide presumed authoritative voices on the issue of climate change. Although, as Merkley and Stecula (2018) contend, these individuals themselves did not necessarily have a significant impact.

In 2000, McCright and Dunlap provided the first detailed examination of ten think tanks, including the Heritage Foundation, which advocated and embraced anti-environmentalism and climate skepticism between 1990 and 1997. These included *The National Center for Policy Analysis (NCPA), Heartland Institute, National Center for Public Policy Research,*

Competitive Enterprise Institute (CEI), Hoover Institution, Marshall Institute, Cato Institute, Citizens for a Sound Economic Foundation (later Citizens for a Sound Economy), Heritage Foundation, American Enterprise Institute (AEI), Reason Public Policy Institute, Foundation for Research on Economics and the Environment, Pacific Research Institute, Claremont Institute. Reflecting changes and developments from international governance strategies following a timeline from the 1992 UN Conference on Environment and Development was the perfect opportunity to mobilize alternative messaging undermining the scientific evidence of climate change, delaying, and obstructing policy efforts.

The (George C) Marshall Institute (1984–2015) which later became the CO_2 Coalition (2015–onwards), consistently opposed the scientific consensus on climate change using discourses of delay. It received funding from conservative donors and fossil industry groups, including Exxon Mobil and the Charles Koch Foundations (Greenpeace, n.d.-d). The organization was founded by William A. Nierenberg, Frederick Seitz, and Robert Jastrow. Frederick Seitz, was previously associated with the Tobacco industry lobby (Hevesi, 2008) and Robert Jastrow would adopt contrarian positions on climate change (e.g., Jastrow et al., 1991). Nierenberg held a PhD in Physics and was a board member of the Science and Environmental Policy Project (SEPP). He also served as a science advisor in the US State Department (Source Watch, n.d.-b). SEPP was initially founded by Fred Singer in 1990 "to challenge government environmental policies based on poor science, The Science and Environmental Policy Project (SEPP) stands for objective science, based on hard evidence" (SEPP, n.d., n.p.). The organization created the Non-governmental International Panel on Climate Change (NIPCC) and was explicitly created to promote opposition to the IPCC. Over time the organization would produce scientific editorials written by several individuals who had disseminated discourses of delay in other outlets concerning the science of climate change and policy (see SEPP, 2011 for example reports).

The Heartland Institute was another headlining organization that elevated countermovement discourse into the public and political environment facilitated by its ties to some Republican politicians and the conservative media echo chamber. Heartland was one of the earliest members of the countermovement network and their actions illustrate the changing discourse over 25 years illustrates the countermovement reacting to broader changes related to "what would work" to challenge legislation and public opinion. Importantly their initial arguments certainly reflected McCright and Dunlap's (2000) initial identification of climate "denial." At the beginning of the movement, anti-science and explicit climate change denial dominated the discourse from the most popular think tanks, often producing pseudo-scientific and conspiratorial positions that McCright and Dunlap labeled: (1) The evidentiary basis of global warming is weak and even wrong, (2) Global warming would be beneficial if it were to occur, (3) Global warming policies would do more harm than good. Over time, these messages have changed as previous climate skepticism was debunked. In 1994, the organization set up a unit focused on climate change and environmental policy. The unit published articles, policy reports, and books, including *Eco-Sanity: A Common-Sense Guide to Environmentalism* to discredit climate science authored by several individuals in its list of "Global Warming Experts" that would critique the science, policy, or environmental activists (e.g. Bast et al., 1994). These individuals wrote op-eds, articles, special reports, made TV appearances, and delivered presentations at events hosted by the institute. The events hosted by the institute also hosted its blog *Climate at a Glance* (n.d.), populated with posts and commentary focused mainly on the science of climate change and climate skepticism. Alongside this, it produced educational material for schools, colleges, and universities.

Another example illustrating the changing discourse overtime was the NCPA which received funding from conservative foundations tied to the Koch brothers (e.g., Brulle, 2014). In their earliest dissemination of discourse, the NCPA utilized the work of Sallie Baliunas. For example, in a dedicated section on Global Warming (NCPA, 1997) as early as 1997 the organization outlined:

> *The Clinton administration has decided to commit the United States to finalizing a treaty in December 1997 that would impose legally binding, internationally enforceable limits on the production of greenhouse gases, primarily carbon dioxide (CO_2). That decision was based on the belief that global warming is significant, that humans are its primary cause and that only immediate government action can avert disaster…Yet there is no scientific consensus that global warming is a problem or that humans are its cause.*

While early iterations combined scientific denial and critiques of activists, in 2003, they created an E-Team which covered general environmental-related issues rather than just isolated comments on the issue of climate change. Furthermore, it helped solidify a sense of authority in producing policy papers and positions (NCPA, 2003). Between 2007 and 2013, the organization maintained some of this critique of the scientific evidence but focused on the economic costs of addressing the consequences through mitigation while recommending adaptation to a warmer world (see Scientific Global Warming Primer, 2007–2013). Moreover, they continued to use the work of individuals that promoted climate skepticism and critiques of climate scientists and activists (e.g., Green & Armstrong, 2008).

Another example is the American Enterprise Institute (AEI), one of the biggest named recipients ($380 million) from fossil fuel interests, promoting several delaying messages. Utilizing the work of those involved in promoting opposition discourse such as Roger Bates' text *Global Warming* (1998) Kenneth Green (e.g., 2009), and articles promoted by the Washington Post (e.g., Wattenberg, 2000), they would also attack Obama's proposed legislation before and following his rise to President. Like other countermovement organizations, the AEI would elevate its propaganda around political opportunities and create targeted responses such as the Copenhagen COP event (e.g., Levy, 2009) and Climategate (e.g., Barone, 2009; Perry, 2009).

A final example is the Heritage Foundation founded in 1973. The organization was a member of the CHC and received grant funding from ExxonMobil and Koch foundations (Greenpeace, n.d.-c). Heritage engaged in several accounts of misinterpreting the evidence on climate change and aligning regulatory action with an additional tax and harming the welfare of the American population. The organization cited several individuals in the organizations during the 1990s (e.g., Antoneilli, 2000; Feulner, 1998; Schaefer et al., 1997a, 1997b), and sponsored countermovement events such as the Heartland Institute Conference on Climate Change (e.g., Climate Investigations, 2017).

Think tanks such as the Cato Institute, Heartland Institute, and the NCPA became instrumental in promoting the fossil fuel industry's interests into the public and political realm. As a result, conservative foundations and fossil fuel corporations could prioritize spending—outside of PAC contributions—outsourcing propaganda that members of the Republican Party, media commentators, and the public could use. While there were no one-size-fits regarding messaging strategies and changes over time by think tanks, their actions significantly increased opportunities to influence legislation and public opinion on the issue of climate change.

Helping produce a cohesive stream of delaying discourses promulgated by US think tanks would be existing networks of organizations. One such network was the State Policy Network (SPN). The SPN became a central hub where countermovement organizations operated under one banner and explored various policy issues. This also became an opportunity to diffuse positions on climate change into state-level policy which were in the interests of the fossil fuel industry. While much of the CCCM may have focused on federal legislation, the SPN was able to diffuse these ideas into and across specific states (Basseches et al., 2022). Think tanks in the network included the Cascade Policy Institute, Independence Institute, John Locke Foundation, and Ethan Allen institute, all of which disseminated delaying discourses at some point in their history (See McKie, 2018 for examples of delaying discourse). Originally called the Madison Group, the SPN received funding from familiar conservative foundations and fossil interests like others in the movement, including $66,351 from the Charles G. Koch Charitable Foundation between 2002 and 2014 (See Source Watch, n.d.-d.).

Similarly, the Atlas Economic Research Foundation, herein Atlas, was an umbrella organization for libertarian and free-market groups across all continents. Previously an ally to the Tobacco industry (Smith et al., 2017), Atlas provided a platform for both neoliberal ideas and delaying discourses to diffuse through the network, alongside operating as a funder to think tanks and we can speculate CCCM activities. For instance, the organization co-sponsored several Heartland Institute climate change conferences (discussed below). It was the recipient of donations from foundations including the John Templeton Foundation (total $9,669,538) and the Dunn Foundation for the Advancement of Right Thinking ($340,000), Donors Trust ($3,003,040), Exxon Mobil ($1,082,500) and Charles Koch Foundation ($595,369) to name a few (see also DeSmog, n.d.-a.).

As well as a recipient of donations, Atlas provided funding to a wide range of conservative think tanks that promoted work encompassing discourses of delay. These include the libertarian Mises Institute (see Chapter 3), the Reason Foundation, and the Acton Institute for the Study of Religion and Liberty. For example, The Reason Foundation was a co-sponsor of the Heartland Institute's International Conference on Climate Change, and a member of the countermovement organization the Civil Society of Coalition on Climate Change (CSCCC) (Chapter 6). Similarly, the Acton Institute for the Study of Religion and Liberty was a free market, religiously influenced think tank (Acton, n.d.) and it co-sponsored the Heartland Institute's International Conference on Climate Change.

Atlas' board of directors had further connections to the CCCM. These included former Chairman Gerry Ohstrom, a previous director of the think tank Property and Environment Research Center (PERC). Established in 1980, PERC promoted market approaches to environmental problems and received funding from Koch foundations and Exxon Mobil (Source Watch, n.d.-c). In addition, the organization was a member of the CHC and held ties to the George W. Bush administration, where previous Executive Director Terry Anderson was on Bush's presidential campaign's environmental advisory team. Also, PERC was previously involved in the Wise Use Movement (Jacobs, 1995).

Similarly, Lawson Bader was president of the CEI (2012–2015), and Vice President of the Mercatus Center at George Mason University (CEI, n.d.). He later became CEO of Donors' Trust and Donors' Capital while serving on the board of the Atlas Network and SPN (Donors Capital Fund, n.d.). These interlocks between individuals and organizations certainly illustrated the successful infrastructure in place to facilitate the diffusion of the same messaging and the opportunity for multiple funding streams to elevate countermovement opposition.

2.4 The Republican Party and Success in a Legislative Context (2001–2008)

Historically, opposition to climate change legislation followed the Conservative movement's anti-environmentalism of the Wise Use Movement. Therefore, the receptivity of Republican politicians to obstruct climate policy is connected to the party's values. In short, the Republican party represented a series of beliefs that justified an existing social order that was somehow challenged. In this context, these beliefs centered on (1) fiscal conservativism and the reduced or limited role of the state needed to be preserved. (2) Social conservativism particularly traditional family values which needed to be preserved. And (3) reduced government intervention regarding environmental issues mainly in response to the growing environmental movement that emerged from the 1970s. Chapter 1 describes that countermovements arise in opposition to activities that challenge social structures (Meyer & Staggenborg, 1996). The Republican party would ultimately align with the CCCM and, as such, would politicize climate change. This is because the fossil fuel economy is intrinsically connected to the existing social order and status quo. Therefore, in the pursuit of protecting and defending the social organization of society, the Republican party played a dominant role in disseminating messages that challenged scientific evidence, making the issue of climate change inherently political. In turn, party politics overwhelmingly entered the debate to undermine efficient and swift climate action, particularly following the election of Bush in 2000.

Between the Rio Conference and the George W. Bush administration, the CCCM would undoubtedly influence the political discourse on climate change in the USA. Its' evident impact is reflected in the Bush administration's rejection of the Kyoto Protocol and the failure to implement appropriate action to address the climate crisis. Moreover, as evidenced above, the connections between CCCM organizations and the Republican Party were important for advancing fossil fuel industry interests during this period when the Republicans dominated the house and senate for most of his tenure. Thus, the success of the CCCM in a legislative context was intrinsically linked to its influence over predominantly—although not exclusively—Republican party congressional members.

One way in which the Republican Party had success in obstructing climate action to protect the interests of the fossil fuel industry is through Political Action Committees (PACs). PACs acted as the vehicle for fossil fuel corporations and countermovement actors to influence and shape environmental policy in the USA. Briefly, PACs were first introduced into the political sphere in 1943, providing unions with the ability to contribute to political candidates indirectly, which previously had been prohibited. Over time, PACs became the optimal way to leverage political actors, and the 2010 Citizens United v Federal Election Campaign (FEC) Act made it easier to directly influence policies (see above). Resultant amendments permitted the fossil industry and CCCM organization to leverage their interests and shape environmental policy with supportive PAC lobbying activities which recruited Republican party members (Ard et al., 2017). Since the 1990s, the top contributors to the Oil and Gas industry PACs since the 1990 included but were not limited to Enron Corp, Koch Industries, Chevron Corp, and Exxon. For instance, between 1990 and 2021, ExxonMobil contributed \$25,414,163 to congressional candidates—predominantly to Republican members—and between 1998 and 2021, \$286,142,742 in lobbying efforts. Since 1989–2021, British Petroleum donated a total of \$6.2 million to PAC's that primarily contributed to support Republican candidates with a peak of \$914,500 in 2000 (Open Secrets, n.d.-b).

One infamous Republican political figure joined forces with the fossil industry and CCCM organizations to directly impact USA climate legislation implementation; Senator James Inhofe (R-OK). Between 2003 and 2008, under the Bush administration, Inhofe was elected as the Chair of the Senate Committee on Environment and Public Works and is famous for his promotion of climate change denial. Between 1989 and 2020, top contributors to Inhofe's campaign included Koch Industries (\$111,150) and the American Consolidated Natural Resources (\$975,000) from individual and PAC donations. PAC money from 1989–2020 from the Oil and Gas industry to Inhofe totaled \$1,094,541 (Open Secrets, n.d.-a). In 2015, a Guardian investigation into PAC funding revealed that Inhofe had received \$10,000 from senior staff from British Petroleum (and supported the 45th President Donald Trump's campaign to withdraw from the Paris Climate Agreement. In addition, Inhofe held close ties with some obstruction organizations including acting as a guest speaker at the Heartland Institute's Fourth International Conference on Climate Change and listed as a *Global Warming Expert* (Heartland Institute, 2010).

Similarly, the Washington Legal Foundation (WLF) represented Inhofe on the issue of Global Warming. It had advocated against environmental legislation and regulation such as the EPA and Clean Air Act. With its non-profit status, the WLF received funding from foundations and corporations consistent with the countermovement including ExxonMobil (1998–2016: \$575,000) and the Koch Family Foundations (1992, 2016, \$12,562) (For more details on funding see DeSmog, n.d.-b).

Also assisting the fossil fuel industry to directly influence legislation was the USA Chamber of Commerce, which between 1989 and 2009 deployed delaying discourses to undermine efforts to introduce climate legislation. Importantly, this may also be a response to the active efforts from as early as the George H. W. Bush administration. In 1989, the Bush administration initiated the US Global Change Research Program resulting in the Global Change Research Act (1990) and in the same year he signed the 1990 Clear Air Act Amendment that would aim to improve standards to address the issues of acid rain, urban air pollution and air emissions. Moreover, it introduced the possibilities of a cap-and-trade system that would infiltrate subsequent administration's policy programs.

In short, the chamber was the largest lobbying group in the USA, acting on behalf of business. As a result, the chamber played a pivotal role in elevating the interests of the countermovement and turning to its member organizations to help inform policy on climate change before informing Congress and the administration (Triedman, 2021). Triedman (2021) describes that from 1989 until 2009 the US Chamber would mobilize discourses of delay into policy proposals and treaties that subsequently would be debated within the House and Senate. Hence, the US Chamber was one of the most significant players in garnering political influence and could impact policy and legislation. To ensure this success, Triedman extends that over 20 years, they disseminated discourses of delay that included:

> *The climate change problem is elusive and the science needs more time to develop, solutions being proposed would be harmful to the American economy and put us at disadvantage to other countries, non-transformative solutions like technological innovation and voluntary action are a viable strategy while scientific and policy issues get resolved.* (p. x)

Not only were the interests of the US Chamber central to disseminating discourses of delay, it played a primary role in faciliating the post-Washington consensus (See Chapter 6). Not only that, the commerce leadership was engaged in lobbying efforts, campaigns, and advertising and was also supportive and involved in the leadership of countermovement coalitions such as the GCC. Significantly, during the H.W. Bush administration (1989–1992), the discourses they promoted included climate skepticism and uncertainty. This was undoubtedly an affinity to the discourses of delay that proved more valuable and acceptable during that period. During the George W. Bush Administration (2000–2008), skepticism and uncertainty remained in the discourse, before resorting to the argument that climate policy threatens fundamental livelihoods and living standards. Simultaneously, they added technological optimism into their discourse, or technological determinism, claiming technology will develop, progress, and reverse emissions.

2.5 The Obama Years: (2008–2016)

The election of democratic 44th president of the US Barack Obama, presented an optimal opportunity to rectify the lame and backward policies of the previous Bush administration. However, this was not to be the case with a legislative environment that was hostile to any clear actions (Kincaid & Roberts, 2013). The clearest example of this failure came with the administration's attempt to introduce a cap-and-trade bill. The Waxman Markey Bill: The American Clean Energy and Security Act, passed in the House before failing to reach and be confirmed by the Senate in 2010, could have been one of the most successful cases of environmental and climate-related legislation in USA history (Contestabile, 2019). However, the influence of lobbying by climate obstruction actors and the particular role of CCCM opposition undermined progress. At this time, countermovement organizations actively campaigned to undermine the Bill. For instance, the National Association for Manufacturers (NAM) developed a multimillion-dollar campaign against the Bill, claiming the bill would have "devastating impacts on the economy" (n.p.). Similarly, the Heritage Foundation disseminated a policy analysis on the implications of the Bill, emphasizing the social and economic costs of such an approach (Paliewicz & McHendry, 2017).

The WFA and several other trade associations and corporations funded, supported, and elevated their campaigns with the American Legislative Exchange Council (ALEC). ALEC claimed to be a "nonpartisan, voluntary membership organization of state legislators dedicated to the principles of limited government, free markets and federalism." (ALEC, n.d., n.p.). Together with largely conservative state legislators and private sector interests, ALEC drafted model legislation to be shared among government members. However, a non-governmental organization (NGO) and investigative journalism campaign revealed ALEC's significant ties to the fossil fuel industry and other corporate interests. The Nation's report revealed the organization had created bills with "Koch DNA" and were simply attempts to mainstream these ideas and protect the wealth of Koch investments, right-wing organizations, and legally binding free market fundamentalism (Graves, 2011).

Despite these revelations, ALEC persisted in engaging in an active campaign to challenge several of Obama's attempts to take progressive action on climate. For example, in 2012, the EEI sponsored the ALEC event on States and Nation Policy Summit and the ALEC Energy, Environment and Agriculture Task Force Reception in 2013. This Task Force produced climate and environment-specific model bills allowing corporations to help reshape the state legislature. For instance, it created the model Bills "The Resolution in Opposition to a Carbon Tax" in 2013, a free market bill for renewable energy, "Market-Power Renewables Act" (2013) and the "Resolution Concerning EPA proposed GHG standards for New and Existing Fossil Fuel Power Plants in 2013." In turn, ALEC continued to use this funding to design and model bills to weaken climate and other environmental-related regulations, particularly at the State level. Consequently, ALEC remained one of the most influential trade/industry-related associations that negatively impact climate-related policy (Influence Map, n.d.).

The alignment of CCCM organizations with the Tea Party further tempered attempts for bold climate ambitions by the Obama administration. Driving forward conservative principles of limited government intervention and restraining taxation, Tea Party members and advocates presented a more rigid argument for increasing and expanding marketization and adopting a fiscally conservative ideological series of beliefs that were strongly tied to racial identities and libertarian utopian thought (Arceneauz & Nicolson, 2012). This economic predisposition connected to the positions on climate change held by some Republican congress people, demonstrating a hostility towards climate-related legislation by creating an "environment protection" versus "economic growth" debate. Significantly, Tea Party political activism and PR campaigns, including utilizing Fox News (Dineen, 2011) provided a perfect opportunity for the Koch brothers—life-long libertarians—to further influence American politics and protect their interests relating the environmental and climate-related legislation.

Before Obama's tenure, from 1984 until 2004, the Koch brothers founded the Citizens for a Sound Economy (CSE). Later this organization became FreedomWorks (2003–present), and Americans for Prosperity (AFP) (2003–present) and would directly target the issue of climate change and support attacks on Obama's policies. Both organizations produced several reports and, unsurprisingly, given their libertarian principles, used scaremongering tactics labeling the scientific consensus as a tool to impose negative and harmful government regulation (e.g., Freedom Works, 2001, 2004). Doing so during the Bush administration when policies were already more favorable to their interests, they then elevated their campaign to help undermine reforms that could have emerged under Obama's tenure. Moreover, Kochs' reach extended to the libertarian think tank and member of the CHC the CEI. The CEI took pride in their "instrumental fighting decades of climate alarmism and anti-energy policies that construct energy supplies, raise prices, and stoke unjustified pessimism about human adaptive capabilities" (n.p., n.d.). The CEI would later play a role in the Tea Party's disruption of the Obama Cap-and-Trade Bill, with the CHC Director Myron Ebell claiming in an interview that: "we turned it into a 'cap and tax' and we turned that into an epithet" (Broder, 2010, n.p.).[1] Similarly, the AFP mobilized the marginalized element of the political right into the mainstream, advocating for the fatalistic embracing of free markets and limited government, and in 2008 began an AstroTurf campaign called "Hot Air Tour" hosting events that focused on opposition to climate-related legislation (Greenpeace, n.d.-a).

The ideological thinking of these libertarian actors was shared with some Republican congressmen. For example, Rand Paul (K-RP) considered himself a Tea party follower. He used arguments advocating small-government intervention on policies at a national level, rather than related international efforts to address climate change, with a specific focus on "jobs and local economies" related to the coal business in Kentucky. Since 2011, Paul has been on the side of anti-environmental vote choices (LCV, n.d.-b). In 2011, he stood on the Senate floor arguing for a bill to prevent EPA from regulating carbon emissions, including a pushing to approve the Keystone XL oil pipeline. That same year he supported the Global Warming Pollution Senate Roll Call Vote 54 amendment presented by Senate minority leader Mitch McConnell blocking the EPA from reducing GHG emissions under the Clean Air Act. In the end, however, this bill was unsuccessful. Likewise, the libertarian-leaning Mike Lee (R-UT) who received PAC donations from among others, the oil and gas sector consistently made anti-environmental votes on the senate floor (LCV, n.d.-a.). Lee supported Global Warming Pollution Senate Roll Call Vote 54 amendment in 2011. Then in 2014, he voted for Senator Inhofe's 2014 amendment to the S.Con.Res 8 regarding an indiscriminate funding cut for federal agencies to curb GHG. And in 2016 voted to support Senator Barrasso's (R-WY) amendment to the S. 2012, the Energy Policy

[1] Myron Ebell later went onto the Trump EPA transition team in 2016.

Modernization Act of 2015 which would promote accelerated deadlines on natural gas pipelines on federal and tribal lands.

One characteristic unique in its magnitude and construction to the US CCCM was how organizations and individuals crafted a relationship between libertarian economic ideas articulated by the Tea Party and advocates within the evangelical Christian movement. While not all evangelical Christians promote climate skepticism, nor do all conservatives follow a libertarian ideology, a range of actors including the Koch brothers, the CEI, and other associated think tanks and front groups operationalized to reach a broader audience and increase support for the Republican party and simultaniously less progressive climate related legislation. Deckman et al. (2017) report on the strategic role of the Republican Party's alliance of fiscal and social conservatives, with a disproportionate number of those living in the American South including groups of white evangelical Protestants who became a solid voting bloc for Republicans. They note a cultural component that bonded ties between this white evangelical group and Tea party economic thought.

Some organizations that would veer into conversations on climate change and impart discourses of delay had more direct ties than others to the Republican Party, and appeared more fiscally and socially conservative, libertarian, and aligned with the Christian right. For instance, the Christian rights group, the Family Research Council, drew on socially conservative cultural values and promoted economic policies that focused on the limited role of government on business. On their initial venture in the climate change debate, one of the earliest statements in 1997 written by Gary Bauer, then president of the organization claimed:

> *This week's global warming conference in Kyoto, Japan will focus on greenhouse gases. But it looks like the real hot air will come from the recommendations the bureaucrats will make to deal with the issue…Don't get me wrong, I support reasonable conservation. I take seriously man's responsibility to care for the world God created. But we should be careful of so-called science that's used to promote a big government agenda.* (1998, n.p.)

While they may have changed over time to recognize the climate crisis, in the early stages they disseminated critiques and attacks against government intervention aligning with the narratives of a conservative ideological agenda that purposefully focuses on a response to the Kyoto Protocol.

Similarly, The Eagle Forum, also a member of the CHC, released statements written by its president Phyllis Schlafly on the Kyoto Protocol and continued to focus its attention on international governance and energy independence and sovereignty:

> *…The evidence to support this theory is so unclear, inconclusive and contradictory that it cannot be dignified by the term science…Clinton can't control global climate any more than King Canute could hold back the tides. But the Kyoto treaty can change the temperature in your home if you can't afford the high energy taxes to keep it warm in the winter and cool in the summer.*

Schlafly was a well-known conservative activist and attorney who had repeatedly reported on and participated in events that disputed the scientific consensus and approaches to climate legislation. (See Schlafly comments and reports at Eagle Forum, n.d.). Importantly, her voice as a conservative anti-feminist activist helped align socially conservative values and libertarian and neoliberal economic thoughts promoting arguments for reducing red tape and regulation (i.e., reducing regulations on businesses).

The Independent Women's Forum (IWF) was another traditional conservative and libertarian cross-over organization. Nancy Pfotenhauer initially ran the organization. She was a former chief lobbyist for Koch industries (1996–2001), and the organization received $844,155 from Koch-related donations between 1997 and 2014 (Source Watch, n.d.-a). Employing traditional conservative values, with women's work, personal liberty, and free markets arguably lauding "feminism" outputs posted by the organization and written by a variety of individuals consistently spread misinformation about climate change. For example, in 2013 author Lukas published a piece on global warming alarmism claiming:

> *The global warming movement may be less a scientific endeavor and more a political game plan for the redistribution of wealth, higher taxation and government regulation, severe limits on oil and gas production, and restrictions on personal freedoms. Perhaps the alarmism isn't a delusion but a tactic.* (n.p.)

Similarly, in an article by Gunlock (2013) they cited the work of James Delingpole outlining that results of the then IPPC report did not "match up" with the arguments presented by environmentalists. Finally, they further critiqued the educational programs aim to extend climate change education into the mainstream curriculum. Again, referencing those who had challenged the consensus on climate change, proposed policies, or critiqued activists, the article critiques:

> *In spite of promises not to politicize the issues, and assurances that the "science" is settled (ah, it's not), it's more likely that children will get a heaping helping of the stuff those of us of a certain age got when we were in school—but worse.* (n.p.)

The fact that there is a critique of grant funded education programs to support the consensus starkly contrasts the funded education opportunities promoted by fossil fuel companies that extended into education programs in the USA (e.g., Gruenewald & Manteaw, 2007).

Lastly, the Cornwall Alliance exemplified the connections between evangelical Christians and countermovement opposition. The Cornwall Alliance was "a network of evangelical Christian scholars–mostly natural scientists, economists, policy experts, theologians, philosophers, and religious leaders" (n.d.-a, n.p.). The organization connected its religious position with an economic philosophy associated with private property rights, "private property rights, entrepreneurship, free trade, limited government, the rule of law, and access to abundant, affordable, reliable energy." President and founder of the organization was E. Calvin Beisner who composed the 1999 *Cornwall Declaration on Environmental Stewardship*. Beisner would also participate in the Heartland Institutes climate change conferences and be awarded the Heritage Foundation's "*Outstanding Spokesman for Faith, Science, and Stewardship Award*" (n.d.-b, n.p.). The Cornwall Declaration on Environmental Stewardship was the first of several self-entitled "*landmark*" documents produced by the organization (n.d.-c, n.p.). These documents would frame human's use of nature as a representation of "God's image, to add to the earth's abundance" (2006, n.p.). These arguments are tied to Lamb et al.'s (2020) "surrender" delaying discourses while simultaneously deploying technological innovations and "non-transformative" solutions.

Over time, the alliance would promote alternative positions to address the climate crisis. Still, attention would ramp up during Obama's tenure and the progressive action the administration at least on the surface proposed. Using political opportunities such as Climategate (e.g., Beisner, 2014) and critiquing democrats' program as a *green agenda* and Marxism in disguise (Gregory, 2012) the organization aligned the Christian right, conservativism, and economic liberalism to discredit climate and environment-related policy during the period of a democratic president. Describing "a hidden agenda," and sharing their concerns that American citizens will fall victim to federal taxes and regulations, the organization combined the issue of climate change into broader political discussions.

2.6 Conclusion

The USA was the birthplace of the organized opposition that emerged to challenge environmental legislation and climate action. Its roots stemmed from the purposeful consolidation of an action plan by the fossil fuel industry and vested interests. Using money to build a network of think tanks, PAC contributions to politicians, and political parties the fossil fuel industry undoubtedly delayed swift and significant action on climate change. At the same time, it remained one of the biggest contributors to GHG emissions. A network of think tanks, coalitions, trade associations, and an accompanying media echo chamber ensured that delaying discourses would penetrate politics and the public minds, increase political partisanship, and lead to political delay. If it were not for this organized campaign, countermovement opposition organizations across other countries may not have had the opportunity to emerge and garner success.

References

Acton Institute. (n.d.). Mission. https://acton.org/about/mission. Accessed 13 Oct 2022.

American Enterprise Institute. (n.d.). Global warming by Roger Bates. https://www.aei.org/research-products/book/global-warming-2/. Accessed 13 Oct 2022.

American Legislative Council. (n.d.). About. https://alec.org/about/. Accessed 21 March 2022.

Antoneilli, A. (2000). Global warming and hot air. https://www.heritage.org/environment/commentary/global-warming-and-hot-air. Accessed 13 Oct 2022.

Antonio, R. J., & Brulle, R. J. (2011). The unbearable Lightness of politics: Climate change Denial and political polarization. *The Sociological Quarterly, 52*(2), 195–202. https://doi.org/10.1111/j.1533-8525.2011.01199.x

Arceneauz, K., & Nicolson, S. P. (2012). Who wants to have a tea party? The who, what, and why of the Tea Party movement. *Political Science & Politics, 45*(4), 700–710. https://doi.org/10.1017/S1049096512000741

Ard, K., Garcia, N., & Kelly, P. (2017). Another avenue of action: An examination of climate change countermovement industries' use of PAC donations and their relationship to Congressional voting over time. *Environmental Politics, 26*(6), 1107–1131. https://doi.org/10.1080/09644016.2017.1366291

Bannerjee, N. (2015, December 22). Exxon's Oil Industry peers knew about climate dangers in the 1970s, too inside climate news. https://insideclimatenews.org/news/22122015/exxon-mobil-oil-industry-peers-knew-about-climate-change-dangers-1970s-american-petroleum-institute-api-shell-chevron-texaco. Accessed 28 Nov 2022.

Barone, M. (2009). Global warming consensus. https://www.aei.org/articles/global-warming-consensus/. Accessed 28 Nov 2022.

Basseches, J. A., Bromley-Trujillo, R., Boykoff, M. T., Culhane, T., Hall, G., Healy, N., Hess, D. J., Hsu, D., Krause, R. M., Prechel, H., & Roberts, J. T. (2022). Climate policy conflict in the US states: A critical review and way forward. *Climatic Change, 170*(3), 1–24.

Bast, J., Hill, P.J., & Rue, R. (1994). *Eco-sanity: A common-sense guide to environmentalism*. Madison Books Inc.

Bauer, G. (1997, December 2). Today's radio commentary. https://web.archive.org/web/19990422085516/http://www.frc.org/radio/c97l02.html. Accessed 21 March 2022.

Beisner, E. C. (2014). Three great legacies of climategate. https://cornwallalliance.org/?s=climate+gate. Accessed 21 March 2022.

Brick, P. (1995). Determined opposition: The wise use movement challenges environmentalism. *Environment: Science and Policy for Sustainable Development, 37*(8): 17–42 https://doi.org/10.1080/00139157.1995.9930953

Broder, J. M. (2010, March 25). Cap and trade loses its standing in energy policy crisis. https://www.nytimes.com/2010/03/26/science/earth/26climate.html. Accessed 21 March 2022.

Brulle, R. J. (2014). Institutionalizing delay: Foundation funding and the creation of U.S. climate change countermovement organizations. *Climactic Change, 122*, 681–694. https://doi.org/10.1007/s10584-013-1018-7

Brulle, R. J., Hall, G., Loy, L., & Schell-Smith, K. (2021). Obstructing action: Foundation funding and US climate change countermovement organizations. *Climatic Change, 166*(1), 1–7.

Carmichael, J. T., Brulle, R. J., & Huxster, J. K. (2017). The great divide: Understanding the role of media and other drivers of the partisan divide in public concern over climate change in the USA, 2001–2014. *Climatic Change, 141*(4), 599–612.

Cato Institute. (2009a). Ad campaign takes on global warming myths. https://www.cato.org/policy-report/may/june-2009a/ad-campaign-takes-global-warming-myths. Accessed 28 Nov 2022.

Cato Institute. (2009b, January 28). Patrick J. Michaels debates global warming on FOX News. https://www.cato.org/multimedia/media-highlights-tv/patrick-j-michaels-debates-global-warming-fox-news. Accessed 28 Nov 2022.

Cato Institute. (n.d.). About. https://www.cato.org/about. Accessed 21 Mar 2022.

Center for International Environmental Law. (2016). Smoke and fumes. https://www.smokeandfumes.org/documents. Accessed 28 Nov 2022.

Climate at a Glance. (n.d.). Homepage. https://climateataglance.com

Climate Files. (n.d.). 1998 American Petroleum Institute global climate science communications team action plan. https://www.climatefiles.com/trade-group/american-petroleum-institute/1998-global-climate-science-communications-team-action-plan/. Accessed 21 Mar 2022.

Climate Investigations. (2017, March 24). Who is paying for Heartland Institute climate Denial-Palooza? https://climateinvestigations.org/who-is-paying-for-heartland-institute-climate-denial-palooza/. Accessed 24 Nov 2022.

Climate Investigations. (n.d.). Consumer alert. www.climateinvestigations.org/consumer-alert/. Accessed 21 Mar 2022.

Competitive Enterprise Institute. (n.d.). Energy and environment. https://cei.org/issues/energy-and-environment/. Accessed 28 Nov 2022.

Competitive Enterprise Institute (n.d.). Lawson Bader. https://cei.org/experts/lawson-bader/. Accessed 24 Mar 2022.

Contestabile, M. (2019). Lobbying impacts. *Nature Sustainability, 2*(7), 541–541.

Cooler Heads Coalition. (n.d.). About. http://www.globalwarming.org/about/. Accessed 21 Mar 2023.

Cornwall Alliance for the Stewardship of Creation. (2006). The Cornwall declaration on environmental stewardship. https://cornwallalliance.org/wp-content/uploads/2014/04/the-cornwall-declaration-on-environmental-stewardship.pdf. Accessed 28 Nov 2022.

Cornwall Alliance. (n.d.-a). About. https://cornwallalliance.org/about/. Accessed 24 Nov 2022.

Cornwall Alliance. (n.d.-b). Who we are. https://cornwallalliance.org/about/who-we-are. Accessed 24 Nov 2022.

Cornwall Alliance. (n.d.-c). Landmark documents. https://cornwallalliance.org/landmark-documents/. Accessed 24 Nov 2022.

Cornwall Alliance. (n.d.-d). The Cornwall declaration on environmental stewardship. https://cornwallalliance.org/wp-content/uploads/2014/04/the-cornwall-declaration-on-environmental-stewardship.pdf. Accessed 24 Nov 2022.

Deckman, M., Cox, D., Jones, R., & Cooper, B. (2017). Faith and the free market: Evangelicals, the tea party, and economic attitudes. *Politics and Religion, 10*(1), 82–110.

DeSmog. (n.d.-a). Atlas Economic Research Foundation. https://www.desmog.com/atlas-economic-research-foundation. Accessed 21 Mar 2022.

DeSmog. (n.d.-b). Washington Legal Foundation. https://www.desmog.com/washington-legal-foundation/. Accessed 21 Mar 2022.

Dickson. (2007). http://129.174.21.13/bitstream/handle/1920/6500/12-97-1-PB.pdf?sequence=1&isAllowed=y

Dineen, K. (2011). Reading the tea leaves: The Tea Party movement, the conservative establishment and the collapse of climate change legislation. https://dspace.mit.edu/handle/1721.1/66804. Accessed 21 Mar 2022.

Donors Capital Fund. (n.d.). Lawson R. Bader. http://donorscapitalfund.org/AboutUs/LawsonRBader.aspx. Accessed 21 Mar 2022.

Eagle Forum. (n.d.). Global warming. https://eagleforum.org/topics/global-warming.html. Accessed 21 Mar 2022.

Feulner, E. J. (1998). No evidence for global warming. https://www.heritage.org/environment/commentary/no-evidence-global-warming. Accessed 21 Mar 2022.

Franta, B. (2021). Weaponizing economics: Big oil, economic consultants, and climate policy delay. *Environmental Politics*. https://doi.org/10.1080/09644016.2021.1947636

Freedom Works. (2001). President Bush: Standing firm to protect people and the planet. https://www.freedomworks.org/press/president-bush-standing-firm-to-protect-people-and/. Accessed 21 Mar 2022.

Freedom Works. (2004). What's the state of global warming restrictions. https://www.freedomworks.org/press/whats-the-state-of-global-warming-restrictions/. Accessed 21 Mar 2022.

Graves, L. (2011, July 12). ALEC exposed: The Koch connection. *The Nation*. https://www.thenation.com/article/archive/alec-exposed-koch-connection/. Accessed 21 Mar 2022.

Green, K. C. (2009). The industrialized world and global warming: What Difference will significant reductions in greenhouse gases make? *AEI Ideas*. https://www.aei.org/economics/us-economy/the-industrialized-world-and-global-warming-what-difference-will-significant-reductions-in-greenhouse-gases-make/. Accessed 21 Mar 2022.

Green, K. C., & Armstrong, J. S. (2008). *Global warming: Experts' opinions versus scientific forecasts*. National Center for Policy Analysis Policy Report (308).

Greening Earth Society. (n.d.). About us. https://web.archive.org/web/20050308202842/http://www.greeningearthsociety.org/about.html Accessed 21 Mar 2022.

Greenpeace. (n.d.-a). Americans for Prosperity Foundation (AFP) Koch Industries Climate Denial Front Group. https://www.greenpeace.org/usa/fighting-climate-chaos/climate-deniers/front-groups/americans-for-prosperity-foundation-afp/. Accessed 21 Mar 2022.

Greenpeace. (n.d.-b). Atlas Network: Koch Industries Climate Denial Front Group. https://www.greenpeace.org/usa/fighting-climate-chaos/climate-deniers/front-groups/atlas-economic-research-foundation/). Accessed 21 Mar 2022a.

Greenpeace. (n.d.-c). Heritage Foundation. https://www.google.com/search?q=heritage+foundation+greenpeace+funding&oq=heritage+foundation+greenpeace+funding&aqs=chrome..69i57j0i546l3j69i64.5257j0j7&sourceid=chrome&ie=UTF-8. Accessed 21 Mar 2022b.

Greenpeace. (n.d.-d). George C. Marshall Institute. https://www.greenpeace.org/usa/fighting-climate-chaos/climate-deniers/front-groups/george-c-marshall-institute/. Accessed 21 Mar 2022c.

Gregory, D. (2012). Brian Sussman's eco-tyranny exposes marxist roots of environmentalism. https://cornwallalliance.org/2012/05/brian-sussmans-eco-tyranny-exposes-marxist-roots-of-environmentalism/. Accessed 21 Mar 2022.

Gruenewald, D. A., & Manteaw, B. O. (2007). Oil and water still: How no child left behind limits and distorts environmental education in US schools. *Environmental Education Research, 13*(2), 171–188.

Gunlock, J. (2013). What's actually scary? Environmentalists with time on their hands. https://www.iwf.org/2013/09/27/whats-actually-scary-environmentalists-with-time-on-their-hands/. Accessed 21 Mar 2022.

Haddad, A. (2008). The problem with a 'cap-and-trade' system. https://www.aei.org/articles/the-problem-with-a-cap-and-trade-system-2/. Accessed 21 Mar 2022.

Heartland Institute. (2010). Fourth international conference on climate change. Program. https://web.archive.org/web/20140725120114/http://heartland.org/media-library/pdfs/ICCCs/ICCC4.pdf. Accessed 22 Nov 2022.

Hevesi, D. (2008, March 6). Frederick Seitz, Physicist who led skeptics of global warming, Dies at 96. *The New York Times*. https://www.nytimes.com/2008/03/06/us/06seitz.html. Accessed 21 Mar 2022.

IMDB. (n.d.). Myron Ebell. https://www.imdb.com/name/nm3052169/. Accessed 2 Sep 2022.

Influence Map. (n.d.). American Legislative Exchange Council. https://lobbymap.org/influencer/American-Legislative-Exchange-Council-ALEC. Accessed 2 Sep 2022.

Inside Climate (n.d.). Global climate science communications plan. https://insideclimatenews.org/wp-content/uploads/2015/12/Global-Climate-Science-Communications-Plan-1998.pdf. Accessed 28 Nov 2022.

Jacobs, H. M. (1995). The anti-environmental 'wise use' movement in America. *Land Use Law & Zoning Digest, 47*(2), 3–8. https://doi.org/10.1080/00947598.1995.10394698

Jastrow, R., Nierenberg, W., & Seitz, F. (1991). Global warming: What does the science tell us. *Energy, 16*(11–12), 1331–1345. https://doi.org/10.1016/0360-5442(91)90006-8

Jones, C., Werthman, C., Motley, D., Triedman, C., & Dolan, M. (2019). Countermovement coalitions. Climate denialist organizational profiles. Climate Development Lab. http://www.climatedevlab.brown.edu/uploads/2/8/4/0/28401609/covercountermovementcoalitions.2.2019.pdf. Accessed 21 Mar 2022.

Kincaid, G., & Roberts, J. T. (2013). No talk, some walk: Obama administration first-term rhetoric on climate change and US international climate budget commitments. *Global Environmental Politics, 13*(4), 41–60.

Krosnick, J. A., & MacInnis, B. (2010). *Frequent viewers of Fox News are less likely to accept scientists' views of global warming*. Report for The Woods Institute for the Environment. http://woods.stanford.edu/docs/surveys/Global-Warming-Fox-News.pdf. Accessed 28 Nov 2022.

Lamb, W. F., Mattioli, G., Levi, S., Roberts, J. T., Capstick, S., Creutzig, F., ... & Steinberger, J. K. (2020). Discourses of climate delay. *Global Sustainability, 3*, e17.

LCV. (n.d.-a). Mike Lee scorecard. http://scorecard.lcv.org/moc/mike-lee. Accessed 21 Mar 2022.

LCV. (n.d.-b). Rand Paul Scorecard. https://scorecard.lcv.org/moc/rand-paul. Accessed 21 Mar 2022.

Levitt, J. (2010). Confronting the impact of citizens united. *Yale Law and Policy Review, 29*(1), 217–234.

Levy, P. I. (2009). Calibrating credulity for Copenhagen. https://www.aei.org/articles/calibrating-credulity-for-copenhagen/. Accessed 21 Mar 2022.

Lukas, C. (2013). Global warming alarmism: Delusion or device? https://www.iwf.org/2013/12/04/global-warming-alarmism-delusion-or-device/

McCright, A., & Dunlap, R. E. (2000). Challenging global warming as a social problem: An analysis of the conservative movements counter-claims. *Social Problems, 47*(4), 499–522. https://doi.org/10.1525/sp.2000.47.4.03x0305s

McKie, R. (2018). *Rebranding the climate change countermovement through a criminological and political economic lens* (Doctoral dissertation). Northumbria University. https://nrl.northumbria.ac.uk/id/eprint/33466/. Accessed 28 Nov 2022.

McKnight, D. (2010). Rupert Murdoch's News Corporation: A media institution with a mission. *Historical Journal of Film, Radio and Television, 30*(3), 303–316. https://doi.org/10.1080/01439685.2010.505021

Merkley, E., & Stecula, D. A. (2018). Party elites or manufactured doubt? The informational context of climate change polarization. *Science Communication, 40*(2), 258–274.

Meyer, D. S., & Staggenborg, S. (1996). Movements, countermovements, and the structure of political opportunity. *American Journal of Sociology, 101*(6), 1628–1660.

Mulvey, K., & Shulman, S. (2015). *The climate deception dossier: Internal fossil fuel industry memos reveal decades of corporate disinformation.* Union of Concerned Scientists.

National Center for Policy Analysis. (1997). Global warming: Scientific debate. https://web.archive.org/web/19971120043304/http:/www.public-.policy.org/~ncpa/hotlines/global/gwhot1.html. Accessed 8 Oct 2022.

National Center for Policy Analysis. (2003). E-Team. https://web.archive.org/web/2003a1123093102/http:/eteam.ncpa.org/. Accessed 8 Oct 2022.

Open Secrets. (n.d.-a). James M Inhofe. https://www.opensecrets.org/members-of-congress/james-m-inhofe/industries?cid=N00005582&cycle=CAREER. Accessed 8 Oct 2022.

Open Secrets. (n.d.-b). British Petroleum. https://www.opensecrets.org/orgs/bp/summary?id=D000000091. Accessed 21 Mar 2022.

Paliewicz, N. S., & McHendry, G. F., Jr. (2017). When good arguments do not work: Post-dialectics, argument assemblages, and the networks of climate skepticism. *Argumentation and Advocacy, 53*(4), 287–309.

Peeples, J. A. (2014). Aggressive mimicry: The rhetoric of wise use and the environmental movement. In *The environmental communication yearbook* (pp. 1–17). Routledge.

Perry, M. J. (2009). The greatest scientific scandal of our age. https://www.aei.org/carpe-diem/the-greatest-scientific-scandal-of-our-age/. Accessed 8 Oct 2022.

Schaefer, B., Annett, A., & Antonelli, A. (1997a). Global climate treaty fosters economic impoverishment and endangers U.S. security. https://www.heritage.org/political-process/report/the-road-kyoto-how-the-global-climate-treaty-fosterseconomic. Accessed 21 Mar 2022.

Schaefer, B., Annett, A., & Antonelli, A. (1997b, October 6). The road to Kyoto: How the global climate treaty fosters economic impoverishment and endangers U.S. security. https://www.heritage.org/political-process/report/the-road-kyoto-how-the-global-climate-treaty-fosterseconomic. Accessed 21 Mar 2022.

Schlaffly, P. (1997, November 26). Clinton may get toasted by global warming. https://web.archive.org/web/19980207194838/http://www.eagleforum.org/column/1997/nov97/97-11-26.html. Accessed 21 Mar 2022.
Science and Environmental Policy Project. (2011). 2011 index of editorials. https://www.sepp.org/science-editorials.cfm?whichyear=2011. Accessed 21 Mar 2022.
Science and Environmental Policy Project. (n.d.). Homepage. https://www.sepp.org/. Accessed 21 Mar 2022.
Smith, J., Thompson, S., & Lee, K. (2017). The atlas network: A "strategic ally" of the tobacco industry. *The International Journal of Health Planning and Management, 32*(4), 433–448.
Source Watch. (n.d.-a). Independent Women's Forum. https://www.sourcewatch.org/index.php/Independent_Women's_Forum. Accessed 21 Mar 2022a.
Sourcewatch. (n.d.-b). William A. Nierenberg. https://www.sourcewatch.org/index.php?title=William_A._Nierenberg. Accessed 21 Mar 2022.
Source Watch. (n.d.-c). Property and Environment Research Center. https://www.sourcewatch.org/index.php/Property_and_Environment_Research_Center. Accessed 21 Mar 2022.
Source Watch. (n.d.-d). State policy network. https://www.influencewatch.org/non-profit/state-policy-network/. Accessed 21 Mar 2022.
Triedman, C. (2021). Chamber of obstruction: The shifting discourses on climate change, 1989 Brown University Climate and Development Lab. http://www.climatedevlab.brown.edu/uploads/2/8/4/0/28401609/chamber_of_obstruction_report.pdf. Accessed 21 Mar 2022.
Vidal, J. (2005, June 8). Revealed: How oil giant influenced Bush: White House sought advice from Exxon on Kyoto stance. *The Guardian*. https://www.theguardian.com/news/2005/jun/08/usnews.climatechange#:~:text=%22Potus%20%5Bpresident%20of%20the%20United,which%20was%20dominated%20by%20Exxon. Accessed 24 Jun 2022.
Washington Legal Foundation. (n.d.). Homepage. https://www.wlf.org/#:~:text=Our%20mission%20is%20to%20preserve,and%20the%20rule%20of%20law. Accessed 24 Jun.
Wattenberg, B. (2000). Triumphalism Triumphant. *The Washington Times*. https://www.aei.org/articles/triumphalism-triumphant/. Accessed 24 Jun 2022.

CHAPTER 3

Disseminating Ideas: European Countermovement Activity

3.1 Introduction

The previous chapter outlined the foundations of the CCCM that emerged in the USA. However, it is not the only country where organized countermovement opposition manifested. From the late 1990s, a well-developed, cross-national alliance of European organizations and individuals presented a sustained opposition campaign to challenge climate and environmental legislation within and across European countries. The purpose was to push back against the increasing interest and concern of the public, some politicians, and political parties. Predominantly focusing on Western Europe, although not exclusively, this chapter explores a strategic alliance of think tanks and individuals that operationalized discourses of delay across and beyond Europe, intending to spread ideas and doubt to impact legislation and public opinion. It begins by accounting for the importance of the political economy and the ideological foundations that fostered the conditions for which countermovement opposition has emerged. Directly tying this to the countermovement, it then explores the role of Anthony Fisher who helped craft a network of organizations, specifically a think tank policy nexus, that would ultimately serve to spread neoliberal ideals into politics and policy across countries. What follows is a reflection on how these ideas interlock with the rise of countermovement oppositions in the USA and other countries.

R. E. McKie, *The Climate Change Counter Movement*,
https://doi.org/10.1007/978-3-031-33592-1_3

3.2 Economic Liberalism, Neoliberalism: European Influence on the Climate Change Counter Movement

To best understand this strategic alliance of organizations and individuals that made up the CCCM across Europe, you must reflect on the ideological foundations of some CCCM actors. Through the drive of ideologists to advance neoliberal thinking across countries and direct and indirect connections with the fossil industry, CCCM organizations positioned themselves much like the organized think tank network that emerged from the USA (Plehwe, 2021). This is because central to the CCCM in Europe was the dissemination of ideas, responding to and challenging increasing collective action associated with environmental movements, and a European bloc of coordinated environmental and climate legislation and governance.

As Plehwe (2021) notes:

> *all kinds of fossil energy-related interest groups increasingly felt the need to act to prevent policy choices considered harmful. The prospect of increasing state interventionism at the same time alarmed neoliberals concerned with free-market economics. Once it was no longer possible to exploit academic uncertainty, a diverse climate change policy opposition stepped up efforts to meet overwhelming scientific evidence in the climate sciences with "evidence" generated by dedicated think tank research.* (p. 153)

Think tanks, particularly those in Europe and of focus for much of this chapter, played a central role in advancing liberalism. The aim was to "convert business preferences into public policy …in noisy political conflicts" (Culpepper, 2016, p. 462). One such political conflict was the issue of climate change. Given businesses relied on the advancement of liberalism and circulating "their ideas" to convert into policy, advancing legislation to address the climate crisis at a national and transnational level would ultimately ignite opposition. Moreover, ensuring "debate," "division," and differing perspectives on solutions or non-solutions to the "climate crisis" would be required to muddy the waters on the most effective approach to address the climate crisis.

3.3 Think Tank Opposition: Atlas Network European Quarter

Much like the USA, from the early 2000s, an alliance of think tanks became a central hub to disseminate discourses of delay and exert influence over the political realm. A prominent figure connecting many of these European think tanks was the Atlas Economic Research Foundation, herein Atlas, founded in 1981 by Antony Fisher. Following a failed attempt to reach political status, Fisher pursued a policy project establishing a network of libertarian and free-market think tanks to influence political debate (Frost, 2002). Fisher's classic economic liberalism followed in the footsteps of thinkers including Milton Friedman, Friedrich Hayek, and the Mont Pelerin Society (MPS), using Atlas as a global initiative to mainstream this economic and political thought, spreading "ideas" across Europe and globally.

Before Atlas, in 1955, Fisher founded the UK-based Institute for Economic Affairs (IEA). The IEA played a significant role in elevating Margaret Thatcher and the then UK cabinets' policies, which led to the dismantling of the UK welfare state and trade unions. At the time, neoliberal policies—designed in the interests of elite corporate and state interests, keen on increasing privatization and reduced government intervention—would come to dominant approaches to public policy. This included growing deregulation and privatization of the UK's water, sewage, and oil industries. This signaled that Fisher could successfully promote his economic thought that indirectly impacted policy using the think tank mechanism. Subsequently, the IEA became a central figure in modern UK politics.

Previously a messenger for the tobacco industry (Tobacco Tactics, 2020a), unsurprisingly, the IEA was an early front-runner espousing discourses of delay globally. In 1994, it published the "*Global Warming: Apocalypse or Hot Air?*" authored by *Roger Bates* (previous director of the IEA's Environmental Unit) *and* Julian Morris. Similarly, the organization published *The Political Economy of Climate Change Science: A Discernible Human Influence on Climate Documents* (Bates, 1996). Both pieces contained criticisms of climate science, politicians, and environmentalists. Following, the IEA used events and publications to disseminate delaying messaging, undermining urgent calls to shift away from fossil fuels, and elevating the voices of individual climate contrarians who also appeared in US countermovement organizations. Some of these events and publications coincided with activities in the political realms regarding climate change that these organizations exploited to further their agenda. For instance, in 2009, following the Climategate scandal, the IEA published comments from contrarian Fred Singer in a specific response to this incident (Singer, 2009):

> *The Climategate disclosures over the past few days, consisting of some thousand of emails between a small group of British and US climate scientists, suggest that global warming may be man-made after all – created by a small group of zealous scientists!* (n.p.)

Drawing on critical events such as Climategate emphasized the reactivity of these think tanks. They recognized a political opportunity to strategically deploy discourses of delay in attempts to undermine policy as the issue of climate change gained greater focus both in the political arena and among the public.

Much like CCCM organizations in the USA, the success of the IEA in producing or supporting outputs aimed to create doubt in climate science and proposed policy was the vital role of individual contrarians. For example, contrarian Dr. Robert L. Bradley, once IEA Energy and Climate Change Fellow, founded the USA-based Institute for Energy Research (IER) as well as holding roles at other countermovement organizations, including the Cato Institute and the CEI. Having worked for Enron Corporation, he was another individual who promoted the work of economist Milton Friedman. These positions helped frame the authors' positions on climate change. For instance, in *Climate Alarmism Reconsidered* (2003), Bradley deployed both traditional forms of climate denialism and critiqued government intervention and political power. The critique of government intervention in policy initiatives aligned with the IEA's dialogue on free markets, reduced government intervention, and deregulation.

While unlike the documented evidence of the financial ties to fossil industries and think tanks in the USA, less is known about the IEA and its economic ties to the fossil industry. However, investigative journalists' and NGO indicators illustrated the potential funding ties between British Petroleum (BP) and the IEA. For instance, BP had an annual corporate subscription to the IEA (Corporate Europe Observatory, 2010) and submitted donations to the American base of the IEA, the American

Friends of the Institute for Economic Affairs (De Smog, n.d.). Additionally, an undercover report from Unearthed revealed that back-door events facilitated by the IEA provided opportunities for off-record meetings between corporate donors, the IEA and political officials, including BP (Carter & Ross, 2018). Thus, while only a glimpse into the hidden relationships between think tanks, fossil fuel companies, and government officials, this investigative work revealed direct and indirect ties between CCCM organizations, the fossil industry, and the ability to reach political ministers and candidates. Moreover, coupled with the IEA's relentless attempt to undermine government intervention to decarbonize the economy, there is no doubt that in a battle of ideas on responding to the climate crisis, the IEA was able to convey these messages into UK politics effectively.

The impact of the IEA is formidable in the history of British politics, yet not the only evidence of Fisher's global success. He founded the London-based International Policy Network (IPN) (1971–2011), previously named the Atlas Economic Research Foundation UK. Like Atlas, the IPN was established on the principles of free-market economics and aimed to expand the neoliberal project. Based in both the UK and USA, the organization received funding from Exxon Mobil between 2003 and 2006 ($390,000), the Conservative foundations, and other groups, including Donors Capital Fund ($1,194,850 between 2010 and 2016), Donors Trust ($435,500 between 2004 and 2018), the Sarah Scaife Foundation ($275,000 between 2005 and 2010), and the Pierre F and Enid Goodrich Foundation ($81,000 between 2005 and 2016) (De Smog, n.d.).

Several think tanks members of the IPN disseminated discourses of a delay from the late 1990s. These included naming a few, Timbro (Sweden) (see also Anshelm & Hultman, 2014; Ekberg & Pressfeldt, 2022), The Friedrich Naumann Foundation (Germany) (see Ströher & Peiser, 2005) and the Austrian Economics Center (Austria) (see also Almiron et al., 2020). In addition, the IPN ran several events and published reports on climate change during its operation. These included the "Stormy Weather's" conference on November 6, 2002, promoting the work of climate contrarians, including Robert Balling. The report *Who is afraid of climate change? Perpetuating poverty through doomsday predictions* critiqued by climate scientists and activists. It claimed that addressing climate change would have negative economic impacts around the same time as COP8 held in New Delhi in 2002. In 2003 the organization held a joint event with the Cato Institute (USA); "*Sustainable Development: One Year after the World Summit.*" Sponsoring organizations of this event included the Brazilian Instituto Liberal and Universidade de Candido Mendes, which became hubs for countermovement opposition in South America (Chapter 6). The IPN attempted to diffuse both its neoliberal thinking into a subset of think tanks in Brazil and the South American region alongside its positions on the issue of climate change. The IPN's position as a CCCM organization was affirmed when it became part of the Civil Society Coalition on Climate Change (CSCCC) (Chapter 6).

Julian Morris, the Executive director and founder of the rebranded IPN between 2001 and its demise in 2011 held roles at the IEA, including co-director of the Environment and Technology Program (until 2001) and later a member of the academic advisory board. With relational ties or work promoted by organizations including the Reason Foundation (USA) and the Heartland Institute (USA), Morris' position and influence on climate opposition discourse would diffuse across countermovement organizations in different countries. For instance, in 2008, his commentary and report: "*Which Policy to Address Climate Change*" formed part of a collection of essays by the CSCCC that would then be distributed across organizations globally. In the opening line of the report, Morris opted for more traditional denialism, arguing that:

> *Falsehoods and non-sequiturs abound in discussions of climate change; often they are found together. So, for example, it is frequently asserted that "the science is settled" (a falsehood) and that, therefore, "drastic measures are required to avert catastrophe" (a non sequitur).*

In 2005, while in his role as Professor at the University of Buckingham, Morris also attempted to impact political legislation by submitting a memorandum of written evidence to the UK parliament undermining the science of climate change and highlighting what he indicated were "negative economic impacts" (2005, n.p.). These are only two examples, but this figure has played a central role in advancing the discourses of delay disseminated across countries.

3.4 On Government Doorsteps: 55 Tufton Street

Since the 1990s, alongside the IEA was a well-hidden, influential network of organizations and individuals, who became powerful voices in spreading discourses of delay in the UK to parliament. The most infamous registered charity was the Global Warming Policy Foundation (GWPF) (2009-present). Claiming to "*inform the media, politicians and the public, in a newsworthy way, on the subject in general and on the misinformation to which they are all too frequently being subjected at the present time*" (GWPF, 2009, n.p.) compared to mainstream think tanks, the GWPF looked more like a fringe group following along the lines of USA organization promoted traditional climate denialism. For instance, on its emergence in 2009, its mission was: "*to bring reason, integrity and balance to a debate that has become seriously unbalanced, irrationally alarmist, and all too often depressingly intolerant*" (GWPF, 2009, n.p.)—combining a critique of the science and environmental activists, this all too familiar narrative featured predominantly in the GWPF's discourse until 2018.

Despite its "fringe" appearance, the GWPF was not limited in its reach, evidenced by its continued presence in UK media and politics. Launched and headed by contrarian Nigel Lawson, the GWPF attracted significant attention. In his political career, Lawson joined the Thatcher government to help oversee the free-marketization and laisse-faire transformation of the UK economy, which became the hegemonic global economic model until the financial crash in 2008. Tying his economic thinking to opposition positions on climate change, it was unsurprising that the GWPF would deploy climate denialism, promoting and justifying free market expansion and policies limiting government intervention in businesses.

Noting that climate change was primarily a question of economics in the UK, the ideological-based proposed solutions would align with the legacy of Thatcher's government and politics on the right-hand side of the political spectrum. In turn, similar positions that CCCM organizations would adopt were also evident in the discourse of Conservative politicians such as Boris Johnson, Rt Hon Peter Lilley, and Conservative Party donors. For example, relational ties existed between prominent Conservative donor Michael Hintze, a hedge fund manager and the GWPF, who in 2015 joined the GWPF Board of Trustees (Carbon Brief, 2012). In addition, Peter Lilley had previously vocalized his opposition to climate action in a 2012 report for the organization: "*What is Wrong With Stern?*

The Failings of the Stern Review of the Economics of Climate Change for the organization."

Similarly, while Boris Johnson's public influence on a global scale only cemented itself following his rise to Foreign Secretary (2016–2018) and later Prime Minister (2018–2022), Johnson, on several accounts, promoted discourses of delay. As a columnist at the right-leaning UK newspaper The Telegraph, Johnson wrote several op-eds that critiqued climate science and spread discourses of delay. For instance, in 2000, he described environmentalists as *Green Warrior*s and *green doomsters*. In 2013, he wrote an op-ed titled; "*It's snowing, and it really feels like the start of a mini-ice age: Something is up with our winter weather. Could It be the Sun is having a slow patch?*" In the article, he admitted to liaising with famous British climate skeptic Piers Corbyn, claiming it was time to consult this "learned astrophysicist" before concluding that misguided and misinformed scientific studies should not develop climate policies.

The importance of using the media to disseminate this discourse partly reflects the opposition echo chamber in the USA. However, unlike the USA whereby climate skepticism and discourses of delay were largely promulgated through conservative networks, in the UK, these were often spread in news outlets on both sides of the political spectrum. Painter and Ashe (2012) reviewed the impact of skeptical voices and their presence in the media across six countries between 2007 and 2010. In the case of the UK, they existed in both right and left-leaning news outlets. Importantly, they also show that of the few articles that included climate skepticism, some were also present in outlets on the left-hand side of the political spectrum (e.g., *the Guardian*) to provide "balance" alongside the consensus view. One example from the data was the exposure of the Climategate scandal, a focal point and opportunity for CCCM organizations to operationalize and challenge a consensus view on climate change. It is however important to note that, in media outlets on the right-hand side (e.g., *the Telegraph*), skeptic views were not regularly balanced with the mainstream consensus.

The notion of *balance is bias* is essential here. For example, suppose equal space or time is given to positions that are ultimately not supported by mainstream scientific evidence. In that case, it elevates skeptic voices, legitimizing these views despite the overwhelming evidence that these are unfounded and unsupported. Moreover, in the UK, the topic of climate change was embedded in the ideological project meaning countermovement opposition was not necessarily challenging the science. Instead, the response to climate change spread across ideological lines as if it were an apolitical issue, framed primarily as an economic issue in the media.

The success of the GWPF and other countermovement think tanks in the UK stemmed partly from their geographic location at 55 Tufton Street, London. It was the base for UK CCCM organizations to influence policy indirectly with the libertarian and neoliberal think tanks that disseminated discourses of delay, including Civitas. Civitas was a think tank promoting limited government, personal freedoms, and free enterprise. Its focus was to reduce the impact of EU environmental regulations on the UK and other related policies that would harm the economy. They have published and supported op-eds by climate skeptics including Bjorn Lomborg. It appeared to first discuss the issue of climate change openly in 2010 suggesting:

> *Policies being pursued to combat global warming are weakening the manufacturing sector, undermining our economic recovery and destroying jobs... Imposing costs on high-energy users will drive them overseas where they will continue to produce carbon emissions...But both sides should be able to agree that reducing carbon emissions should not take priority over job creation.* (2010)

This organization was located next to the IEA and the GWPF, on the same street and closely located to Westminster were other CCCM organizations. For example, the Adam Smith Institute, a member of Atlas and the CSCCC (2007–2014), promoted delaying discourses from 2002. The organization consistently critiqued approaches to government policy, particularly those that proposed increased government intervention, regulation, and collective action promoted by environmental activists. For example, in response to the UK Commons vote on the Climate Change Bill, the organization promoted Bowman's piece that cited Julien Morris and the IPN (2008). In 2009 it promoted the work of contrarian Ian Plimer, and in 2010 it promoted a blog by commentator Brandy Patty reflecting on the COP event in Cancun. Much like their USA counterparts, the discourses of delay sustained a critique of government intervention aligning with the ideological trends of the organization as well as explicitly targeting events and periods with heightened engagement on the issue of climate change.

Since 1997, the Centre for Policy Studies (CPS), also a member of Atlas, promoted publications and organized events undermining the urgency for climate action. For example, in 2007 they promoted the report "*Climate Change: A Guide to the Scientific Uncertainties*" by Martin Livermore, former director of the Scientific Alliance, a UK organization (see McKie, 2018 for examples of a discourse of delay from this organization). In 2010, the CPS held a debate involving Tim Yeo (MP) and Nigel Lawson, opposing positions on a UK green economy. The CPS was also a key player in developing Conservative policies during the Thatcher era (e.g., Harris, 1996; James, 1993) and under David Cameron, who later became Prime Minister (Sowel, 2007). This close relationship between the CPS and the UK Conservative Party reflects the similar policies adopted by incoming governments. For instance, climate skepticism was greater among conservative supporters in the UK (e.g., Clements, 2012).

Moreover, despite claiming "green credentials" since the 2010 election under David Cameron's leadership, the Conservative Party, alongside right-wing British press (e.g., *The Telegraph*, *Daily Mail*, *The Spectator*), enhanced partisanship on the issue of climate change. As a result, it could indirectly delay the effective implementation of the policy if at all (Carter & Clements, 2015). This is because central to achieving outcomes for the CCCM was the requirement to influence the UK's political landscape, without which the impact of these organizations would have been negligible at most. That is, CCCM opposition in the UK would not have been successful without spreading "ideas" into policy and public discourse.

The Social Affairs Unit, founded in 1980, is another right-leaning group, much like the IEA counterpart, that disseminated reports or posts that included delaying discourses. For instance, in 2005, it produced a series of articles written by its President and Council member Richard D. North in the lead-up to the G8 Summit, another significant event that the countermovement organizations would exploit. Across this series of articles, the focus of delaying arguments covered issue of social justice and avoiding a green energy transition. In other words, climate action could increase energy poverty. Therefore, there should be limited government interference while the science remained uncertain. North would also become a media fellow for the IEA and claimed he was a: "*climate change skeptic*" (n.p.).[1]

[1] Declaration of interest: In 1996 BP paid for my expenses as I investigated charges that it was behaving badly in Colombia. My conclusion was that it was a responsible firm

Former European Parliament (MEP) member and policy advisor to Thatcher's government (1982–1986), Christopher Monckton held roles in UK and non-UK-based countermovement organizations, including the USA-based Heartland Institute as a policy expert, the Science and Public Policy Institute (SPPI) as a Chief Policy Advisor (Heartland Institute, n.d.), and the Australian Galileo Movement as an Independent Advisor (see Chapter 4). Following Britain's exit from the European Union (BREXIT), he founded Clexit with other well-known individuals that feature in other countermovement organizations or recorded evidence of disseminating delaying discourses, including Vaclav Klaus, Marc Morano, and Willie Soon (Clexit, n.d.). Clexit aimed to:

> *"prevent ratification of the costly and dangerous Paris global warming treaty which is being promoted by the EU and the present US administration..."* claiming to have attracted *"...over 60 well-informed science, business and economic leaders from 16 countries."*

Thus, the organization was not simply focused on the UK. Instead, it aimed to appeal across Europe to convince other organizations to support and adopt positions that promoted the rejection of EU environment and climate-related regulations and broader international governance strategies to address the climate crisis. Reaching a global audience was reflected in expanding the scope of its delaying narratives outside of a UK-centric position. For example, in 2015 the organization stated on its website:

> *For developing countries, the Paris Treaty would deny them the benefits of reliable, low-cost hydrocarbon energy, compelling them to rely on biomass heating and costly weather-dependent and unreliable power supplies, thus prolonging and increasing their dependency on international handouts. They will soon resent being told to remain forever in an energy-deprived wind/solar/wood/bicycle economy.* (2015)

Clexit adopted positions focused on undermining climate action in the UK and Europe and used these same messages when appealing to other

doing good in the country. (Shell had earlier paid my expenses for an investigation of its operations in the Niger delta of Nigeria, which reached similar conclusions.) I often wrote damning BP's "Beyond Petroleum" mantra which distracted its audiences from the firm's real obligation to be competent. https://iea.org.uk/blog/speaking-for-bp-cross-your-fingers-be-patient.

nations. Importantly, these messages could then be reproduced in the voices of global south CCCM organizations or individuals (see also Chapter 6). In short, because some UK countermovement organizations were close to lobbying organizations having direct access to politicians, this created a physical gateway for the interests of the fossil industry to penetrate UK politics. Moreover, it also became clear that individuals in the UK arm of the CCCM would play a vital role in spreading the messages that organizations in other European countries could use.

3.5 Individuals, Government Officials, and the Diffusion of Climate Delay in Europe

In this case, particular individuals, not necessarily defined as climate contrarians, would play key roles in disseminating discourse moving between government roles and think tanks. One such example is in Spain and former Prime Minister Jose Maria Aznar. Aznar moved on from his political position to founding the Foundation for Social Studies and Analysis (FAEs) think tank. Rather than adopt climate skeptic positions, FAEs promoted neoliberal/libertarian policy positions to solve the climate crisis, including tax reductions and market stimulation to improve "competitiveness" in the industry. For example, outputs by members of FAES advocate for reduced regulation in the Spanish electricity market (see Pons-Hernández, 2022). Thus, promoting positions on energy policies based on ideological values underpinned countermovement organizations rather than denying the science of climate change.

A similar example comes from Italy and former Italian Minister Gaetano Quagliariello, a former Italian Conservative party Forza Italia member before forming the "New Center Right" *Nuovo Centrodestra* party in 2013.[2] Quagliariello was President of Fondazione Magna Carta (FMC), a think tank "***dedicated to scientific research on major political issues***," promoting a form of conservative liberalism. Notably, the organization published articles that include those of contrarians such as Bjorn Lomborg (FMC, 2010a) and exploited the Climategate event much like the activities of other CCCM organizations (FMC, 2010b). For example, in 2010, the organization included an editorial that created doubt among the ideas of the climate crisis citing the Oregon Petition (Chapter 6)

[2] The party later dissolved and was replaced by the party Popular Alternative in 2018.

in response to the solidification of the Copenhagen Agreement. The editorial claimed:

> *The Copenhagen agreement foresees 30 billion euros in aid in the three-year period 2010-2012 and another 100 billion by 2020, but these are excessive figures that are likely to fall; the sending of aid will hopefully be linked to monitoring the use that the local ruling classes will make of it. Pragmatism, time and perseverance are the right recipes for tackling the climate issue.*

This position illustrates some interesting discourses, particularly the links to [foreign] aid. Notably, since the beginning of the development of the UNFCCC and subsequent treaties, central is ensuring that those least responsible are compensated. These proposals include financial support for adaptation mechanisms and payment for loss and damage. However, in a report responding to the Copenhagen Agreement, the authors claimed:

> *"It cannot be forgotten that there are countries that use "global warming" in a somewhat rogue way, generating in public opinion a mixture of guilt and resentment that is good for raising the amount of the check detached from the rich countries to the poor ones... A blackmail attitude that in Copenhagen found echo in the proposal of the Argentine prosecutor Antonio Gustavo Gomez to establish an Ecological "Criminal Court against "climate crimes"."*

This example articulates a specific critique of this practice, claiming it was a form of "blackmail" particularly targeted at prosecutor *Antonio Gustavo Gomez* who took significant steps to advance environmental justice campaigns. This particular response directly opposed environmental movements; an essential technique used by CCCM organizations to undermine collective action to environmental problems. In essence, it illustrates how the actions of individuals and organizations embroiled in this network shared tactics with those observed in the US, an activity central to creating consistent messages and creating a global-level opposition movement.

It was also the case that the work of contrarians or those of organizations would be translated and promoted by different organizations across Europe. One such example is the translation of documents written by others by the Bruno Leoni Institute. The Bruno Leoni Institute was an Italian think tank that Almiron and Xifra (2021) note very much mirrored itself on the likes of Cato and the Heritage Foundation and formed part of the Civil Society Coalition on Climate Change (see Chapter 6). Established in 1993, its philosophy is associated with "liberal…marketist" positions (Bruno Leoni Institute, n.d.). This organization would use the translated work of Nigel Lawson (e.g., 2006a, 2006b) and Indur Goklany (2005), particularly closely following its emergence to begin cultivating its positions on the issue of climate change that matched others in the CCCM. The purpose of circulating translated materials certainly emphasizes the important role of sharing ideas while ensuring a consistent and cohesive message that could penetrate the policy sector across different countries.

A similar example is Vaclav Klaus, former Prime Minister of the Czech Republic. Best known as an economist and politician, Klaus, during both his time as prime minister and following, engaged in activities and disseminated discourse that echoes those of other climate contrarians. He shared his positions in mainstream media outlets, including the Financial Times (Klaus, 2007), while also among other CCCM organizations joining the Heartland Institute (Heartland, n.d.) and critiquing environmentalism on his blog (Klaus, 2010). Klaus' work was also translated into Spanish and disseminated by FAEs including the delivery of a presentation on the issue of climate change (El Pais, 2008) demonstrating the particular roles of individuals in helping diffuse the same messaging across countries. It is important to note that Klaus was a significant force in mainstreaming conservative policy in the Czech Republic. Importantly, unlike other Eastern European countries during the late 90s and early 2000, Klaus expanded and ultimately normalized free marketism much like Thatcher in the UK (Hanley, 1999). While there were tensions in the application of Czech Thatcherism and traditional conservative policies in the country, the actions of Klaus mirrored the same activities and ideology of Nigel Lawson from the UK very much so. Whether Klaus' positions impacted policy is up for debate, yet his position and those of others engaged in shifting between government positions and broader policy-related roles would ultimately connect CCCM think tanks and spread ideas to politics and the public.

Individuals took a leading role in expanding the positions of think tanks into governments outside their own country showing the diffusion and interconnectivity of different movement actors across countries. A vital example of this is from Spain. The libertarian think tank the Instituto Juan de Mariana (2005-present), a member of Atlas, directed one of its launch events towards the Kyoto Protocol (Corporate Europe Observatory, 2010). Representatives from other countermovement organizations, including former CEI Senior Fellow and Christopher Horner, were in attendance. Horner maintained a role in the countermovement organization Energy and Environmental Legal Institute (USA). He was given mainstream platforms, including publishing articles in the Washington Times, attended and spoke at several Heartland Institute International Conference on Climate Change and was also a past director of external relations at the European Enterprise Institute; another countermovement organization located in Belgium (McKie, 2018).

Similarly, a founding member economics professor and libertarian, Gabriel Calzada Álvarez, like many contrarians, Álvarez made appearances across several countermovement organizations. He was an advisor to organizations including Ludwig Von Mises Institute (USA) (The Mises Institute, n.d.), and the Center for the New Europe (Belgium) and his work and reports were reproduced or included by others, including the Heartland Institute (Álvarez, 2009). Founded in 1993, The Center for the New Europe was a member of the IPN and again like other CCCM organizations at its core advocated for deregulation, increased freedom of markets and reduced government regulation. The organization held several events over its operational period, including four events on the issue of climate change, particularly during 2004.[3] These four events occurred during the year before the ratification of the Kyoto Protocol signed in 1997. Targeting specific periods due to heightened engagement on climate change is paramount for potentially impacting public opinion or policy decisions (Meyer, 2004). Even simply creating doubt or presenting an alternative vision prevents progress by creating questions, hypotheticals, and reasons for delaying action and intervention.

Diffusing his ideas directly into the political realm, in 2009, Alvarez provided testimony before the USA House Select Committee on Energy Independence and Global Warming in response to Obama's proposed introduction of the Cap-and-Trade Bill. The tenants of Alvarez's testimony called for a restriction on government intervention and subsidies for renewable energy resources sought to delay a renewable energy transition.

[3] Events are as follows led by key climate contrarians: Kyoto Protocol is Dead; good riddance, 27/05/2004, Global Warming: 'The Five Ws' 11/05/2004, Climate Alarmism Reconsidered, 9/11/2004, EU Climate Change Policy: Does it Really Matter if Russia Ratifies Kyoto? 7/07/2004.

This statement to the House Select Committee was directed towards the Obama administration, which had proposed a similar "Spanish Model." Drawing on other discourses of delay, Alvarez stated in the testimony:

> *Deliberately pursuing more expensive and less efficient energy in order to create green jobs has been the source of social harm and net job destruction, and many citizens of a nation are hurt when such policies are pursued.* (n.p.)

While these proposed ideological-oriented positions may be considered "solutions" to the climate crisis, they were often choices based on elite interests. In essence, the proposals promoted the idea of the market solving the problem, in turn, the public will not be disadvantaged (i.e., reduced economic growth) despite the transition from fossil fuels to renewable energy essential components of addressing the climate crisis.

3.6 Dutch Shell, Frits Böttcher, and Public Relations

In Denmark, think tanks, individual climate contrarians, and actors with vested interests became vehicles for disseminating discourses of a delay from the early 2000s. This included the Center for Political Studies (CEPOS), founded in 2004, and a member of Atlas and the defunct IPN. Like other think tanks in the CCCM, the organization promoted personal freedom, less government, and lower taxes. For instance, in 2013, CEPOS promoted the report which claimed: "*We face huge man-made challenges in the coming generation. But climate change is not among the biggest problems…*" In addition, CEPOS promoted research and produced reports to undermine wind energy investment and supportive regulations on a domestic policy level. In 2009, CEPOS and the US-based Institute for Energy Research (IER) commissioned and published the report: *Wind Energy—The Case of Denmark*. This report claimed wind energy in Denmark accounted for about 10% and not 20% of the country's electricity consumption wind energy slowed Denmark's economic growth (CEPOS, 2009). The IER had previously attacked Obama's renewable energy plan. The 2009 study was submitted to the Committee of Health, Education, Labor and Pensions to the US Senate on the issue of labor, working families, and the "American" Dream. Moreover, Greenpeace reported that CEPOS was awarded a $100,000 grant from the Atlas Economic Research Foundation (supported by both the Charles G.

Koch and Claude R. Lambe Foundation), illustrating indirect ties to this extended network of fossil industry groups.

Following the landmark investigation in 2019 by the Platform Authentic Journalism, requests to several ministries, municipalities, and administrative bodies to provide all documents between 2005 and 2019 related to Dutch Shell and its subsidiaries revealed the complex interplay between fossil industries, Denmark, and the delayed action on climate change. This is important, given Shell was integral to the Dutch energy sector and their links with CCCM organizations in the Netherlands from 2010 onwards. Not unlike (university) research institutes in the USA that had been funded by fossil interests and disseminated research utilized by the fossil industry, in the Netherlands, there was also evidence of a similar practice demonstrated in the case of the Rotterdam School of Management (RSM). The RSM had received contracts with Shell and ING, and in a report by Hüzeir and Fraser (2017):

> *Shell, BP, ExxonMobil, Gazprom, and many other fossil fuel energy companies benefit in far-reaching ways from interactions with the business school. This is problematic because these companies rely squarely on continued production and consumption of fossil fuels. They are responsible for unprecedented amounts of greenhouse gas emissions, the dominant cause for recent global warming. RSM's support for their business models renders the faculty complicit in facilitating climate change.* (p. 1)

While investigations on the RSM indicated this funding stream's limited impact (Smaling, 2019), it exposes the potential interaction of university research institutes and the CCCM in the Netherlands.

Similarly, the role of the Dutch Shell fed lobbying and influence on The Hague. Influence Maps (n.d.) highlighted the contradictions in the proposals and the impacts on climate change policy. Dutch Shell provided positive feedback towards regulatory efforts and, at the same time, advanced fossil fuel interests with a particular reliance on promoting clean technologies. This infers that Shell's alleged positive actions to influence climate policy since its early experiences appear to have changed; however, the multinational maintains ties to several countermovement organizations, including the USA Chamber of Commerce, NAM, and the API (2021). Moreover, the ABDUP, one of the oldest lobby clubs in the Netherlands, consisting of AkzoNobel, Shell, DSM, Unilever, and Philips, supported the outcomes of Shell's attempts to undermine climate legislation. This lobbying club reflects the same coalition of the original fossil industry actors identified in the USA that sought to delay climate action and impact public policy. The likes of Shell had already been active in their USA lobbying and donations, and this more recent evidence highlights the transnational nature of their attempts to historically undermine climate legislation.

Part of the success of undermining climate policy more broadly in the Netherlands and Europe between 1989 and 1998 could be attributed to the role of Frits Böttcher a key individual and spokesperson that at least in the case of Shell was regarded as an ally (Beunder, 2020). In 1996, he published the article "*Climate Change: Forcing a Treaty "in the peer-reviewed academic journal Energy and Environment".*" In the paper, he describes the actions of scientists and the development of the IPCC claiming that the story of global action on climate change was a "well-orchestrated campaign" (p. 377). He was a board member for the USA-based countermovement organization Science and Environmental Policy Project (SEPP) and signed the Leipzig Declaration a 1995 statement initially penned by climate skeptic Fred Singer under SEPP that both critiqued the scientific basis alongside challenges to the UNFCCC. The Leipzig declaration was created following a 1995 conference "The Greenhouse Controversy" at the European Academy for Environmental Affairs. A private organization founded by Helmut Metzner, in 1993, the organization would host an event Global Warming: Fact or Fiction. The second conference in 1995 Controversy and Ozone-Problem' led to the creation of the declaration (Sourcewatch, n.d.). While far less is known on the European Academy for Environmental Affairs organization, the Leipzig declaration would be a framework for subsequent declarations and talking points in the early activities of countermovement opposition.

Bottcher later supported the Heidelberg Declaration part of the Dutch Heidelberg Appeal Netherlands Foundation (1992) mainly in response to the convening of the first UNFCCC conference in Rio. The declaration there featured the following claim:

> *an irrational ideology which is opposed to scientific and industrial progress, and impedes economic and social development...the authorities in charge of our planet's destiny against decisions which are supported by pseudo-scientific arguments or false and non-relevant data ...The greatest evils which stalk our Earth are ignorance and oppression, and not Science, Technology and Industry.*

This declaration and proposals from within the CCCM even spread into mainstream media discourse where the declaration was featured in the US Wall Street Journal (CHEMEurope, n.d.). In addition, both the Leipzig and Heidelberg Declaration share familiarities with the work of the Cornwall Alliance and the Evangelical Declaration on Global Warming (Chapter 2).

As Beunder (2020) note, working alongside climate skeptics including Roger Bates and Dr. John Emsley, Böttcher founded the European Science and Environment Forum (ESEF). It was set up from 1994 to 2005 and had previously been part of the Tobacco industry lobbying strategy and network of supporting organizations. The ESEF was mainly concerned with impacting public opinion and accessing media. They sought to provide a platform for scientists who argued they were *not being heard* but had a contribution to climate and/or environmental-related policy. Part of the ESEF's success was the PR support received from APCO Worldwide; a company previously linked to the Advancement of Sound Science Coalition (TASSC) (Chapter 2) and a legacy institution of the Tobacco Industry's campaign to undermine related legislation (1990–2005) (Tobacco Tactics, 2020b). Therefore, activities of Böttcher are significant, connecting some of the mechanisms of the USA countermovement into and across Europe. He was able to help mainstream opposition to reports and feedback from the UNFCCC using significant connections to lobbying groups and political elites, to echo the sentiments of fossil fuel interests that could delay the effective implementation of climate policy across Europe.

3.7 Libertarianism, Political Education, and Dissemination of Discourses of Delay

Similar libertarian forces, like the USA-based CEI, played an active role in Europe, spearheaded by The Mises Institute Network and the Property and Freedom Society to disseminate delaying discourses and potentially influence political policy. Founded in the USA by Lew Rockwell, Murray Rothbard, Burton Blumert, and Henry Hazlitt in 1982, the Mises Network operated in several European countries, including, Spain, Germany, France, Italy, and Poland, to name a few. This network promoted discourses of delay in reports and outputs written by individuals within the network. The Mises network embodies the economic traditions of Ludwig Von Mises, along with others from the Austrian School of Economics, including Fredrick Hayek. Central to the dialogue at the center of the Mises institute network is a “free-market capitalist economy and a private-property order that rejects taxation, monetary debasement, and a coercive state monopoly of protective services” (Mises Institute, USA, n.d.). Thus, these are synonymous with the roots of countermovement opposition. However, this is a recent history, where attention has only significantly turned to the issue of climate change since 2016 (Plehwe et al., 2021).

While it is unclear direct connections between the Mises Network to the fossil industry or its affiliation to a political party, what is important is their role in political education and pursuing the goals of expanding libertarian thinking worldwide. As Bessner (2014, p. 441) reported the “LVMI and guided by Rothbard and Rockwell, promoted a revolutionary strategy centered upon training a cadre of intellectuals to release ordinary Americans’ supposedly ingrained libertarianism.” This dissemination of Ludwig’s ideas into mainstream discourse and educational programs organized and promoted by the Mises network were central to influencing the ideological positions of future and current politicians and public members who may gain access to related material including the media. As such, political education plays a significant role in how individuals who may follow a life in politics or the public more broadly perceive an issue and their subsequent response; such positions were evident in members of the Republican Party (Chapter 2).

Moreover, Mises came from the broader school of Austrian Economics. Importantly this political thought was optimized in the think tank the Austrian Economics Center which had promoted editorials or posts that disseminated delaying discourses (e.g., Tupy, 2017) with a previous scientific board member Nigel Lawson (Austrian Economics Center, n.d.). But why is this important in the context of the CCCM? The thought project, and the spread of ideas under this political and economic philosophy contributed to the broader narratives which promoted and sustained fossil capitalism for so long. In turn, it would become easier to utilize an existing network of ideological aligned organizations to develop a cohesive set of positions on the issue of climate change that could be deployed across countries into policies and the public mind. Notably, outlining and sharing the same ideas builds a cohesive message and strategy that could prove more successful in creating a global impact.

3.8 Consolidating Networks: European Climate Realist Network

The European Climate Realist Network (ECRN) was founded in Denmark in 2018, however, the organizations that made up the network had played a role in the countermovement before its formation (see Table 3.1). The European arm of CFACT (Germany), since its inception in 2004, echoed the climate skepticism from its US-based partner (CFACT). Unsurprisingly, the organization translated and published the work of climate contrarians including Paul Driessen. Also, CFACT Europe founder Holger Truss supported the European Institute for Climate and Energy (EIKE) foundation. EIKE, founded in 2007, was a think tank in Germany dedicated to disseminating oppositional positions on the climate change issue that included both the denial of science, policy-focused responses, and critiques of collective action by environmentalists. EIKE importantly was first and foremost a group to rally public support for a fringe campaign to elevate climate skepticism in its traditional forms in Germany that the public and political parties could adopt. Notably, several members of EIKE were active with the Christian Democratic Union of Germany, the Social Democratic Party of Germany, and the Free Democratic Party (Heartland Institute, n.d.). Eike claimed to be "an association of a growing number of natural sciences, humanities, and economic scientists, engineers, journalists and politicians, the assertion of a 'man-made climate change' is not justified as a science and therefore a fraud against the population" (2015). Since its inception in 2007, it continually promoted traditional skeptic arguments in its discourse. For instance, its earliest manifesto in 2007 claimed:

> *1) The climate cannot be demonstrably influence by man-made CO2 emissions. 2) The scenarios of the future development of the climate derived from climate models are speculative and contradict the climate history. 3) In the history of the earth, there has always been climate change with alternating warm and cold periods. 4) The trace gas co2 does not pollute the atmosphere.* CO^2 *is essential for plant growth and therefore a perquisite for life on earth. 5) We are committed to effective protection of our environment and advocate measures that prevent unnecessary stress on ecosystems. 6) We warn against taking measures that do not benefit our environment and cause economic damage under the guise of a conjured up 'climate catastrophe'.*

Table 3.1 The European Climate Realist Network

Name of organisation	*Country*	*Organizational purpose*
EIKE	Germany	EIKE as a public forum is to present the climate and energy facts without ideology, to organize international scientific congresses, to promote and disseminate publications on climate research, in particular from EIKE members
Science, Climate and Energy	Belgium	*Explain the scientific theories most conveyed in the media and combat received ideas. No theory is perfect and our goal is to present the limits of these theories. To explain scientific theories, especially those generating most of the "buzz" in the media, and overcome stereotypes. No scientific theory is perfect and our aim is to present the limits of the theories*
Climate Realism	Denmark	*Climate realists have a research and knowledge-based approach to the climate challenge, which leads us to the following considerations:* *1. We of course recognize that climate change is taking place and that it is partly man-made* *2. There is great uncertainty about how big climate change is and will be and the UN climate panel, the IPCC, tends to overestimate the influence of humans and the extent of climate change* *3. Climate change has both positive and negative consequences. Until now, they have been net positive for human prosperity and well-being. The IPPC estimates that a temperature increase of up to 2 degrees will affect global GDP by between 0.2 and 2%. Which is not an insurmountable challenge for an increasingly rich world* *4. We therefore warn against any kind of climate policy that makes energy artificially expensive, reduces economic growth and thereby risks becoming many times more expensive than the possible, alleged future consequences of climate change* *5. We do not believe that with today's technology it is possible to replace fossil fuels with anything other than nuclear power. Without the fact that it will lead to a dramatic lowering of the standard of living to the great detriment of especially the poorest people in the world* *6. We must therefore do more research into climate change and its natural and human causes, and into developing new energy sources*

(continued)

Table 3.1 (continued)

Name of organisation	*Country*	*Organizational purpose*
Climate, Environment and Energy	France	*The purpose of the association of climate-realists is to promote an open and free debate on the evolution of the climate and the societal and environmental questions relating to it, by promoting the expression in all its forms of rigorous and argued opinions. It aims to raise citizens' awareness of climate issues and energy policies carried out in the name of the fight against global warming. The association is apolitical and totally free in the expression of its ideas. It strives to disseminate reliable information gathered from serious sources. The position of climate-realists on the evolution of the climate and "climate policies" is summarized in a 4-page brochure accessible by clicking on this link.* https://static.climato-realistes.fr/2018/06/4Pages.pdf
Irish Climate Science Forum	Ireland	*The Irish Climate Science Forum (ICSF) was founded in 2016 and its members include Irish scientists, engineers and other professions. We are committed to identifying and disseminating the latest climate science to all with an open and enquiring mind, driven by the imperative of objectivity without vested interests. We are self-funded through modest personal contributions; to ensure scientific objectivity, we do not accept corporate or sector funding.* https://www.icsf.ie/
Climate Intelligence Foundation—CLINTEL (International—Clintel.org)	The Netherlands	*1. The Foundation tries to communicate objectively and transparently to the general public what facts are available about climate change and climate policy and also where facts turn into assumptions and predictions. 2. The Foundation conducts and stimulates a public debate about this and carries out investigative reporting in this field. 3. The Foundation wants to function as an international meeting place for scientists with different views on climate change and climate policy. 4. The Foundation will also carry out or finance its own scientific research into climate change and climate policy* *CLINTEL wants to take the role of independent 'climate watchdog', both in the field of climate science and climate policy*

(continued)

Table 3.1 (continued)

Name of organisation	*Country*	*Organizational purpose*
Klimarealistene	Norway	*The climate realists represent the climate realist majority in Norway's population (Note 1). We are a party-politically independent organization for us who believe that the climate is dominated by natural variations. We do not agree with the UN climate panel, when they claim that emissions of CO_2 change the climate dramatically. We support scientifically based public information as an important part of democracy* *The realization of our purpose is dependent on financial support from a variety of private contributors through membership and donations*
Klimatsans	Sweden	*Our aim is to spread the facts and explanations behind the media's news. We do that here on the blog, in debate articles and submissions as well as lectures for interested congregations. It will be a counterweight to the many ill-founded disaster alarms, which the media often trumpet with big headlines*
Global Warming Policy Foundation (renamed Net Zero Watch, 2020)	UK	*Net Zero Watch will scrutinize policies, establish what they really cost, determine who will have to pay, and explore affordable alternatives* *Increasingly conversations about climate and energy policy are becoming narrow and polarised. Net Zero Watch is here to give you a clear view of the reality of climate and energy policies and what they mean for you* *We all want to see a cleaner and better environment for future generations, but if climate and energy policies are rushed and their costs and implications not fully thought through, they will almost certainly do more harm than good* *Net Zero Watch is run by the Global Warming Policy Forum*

Moreover, in 2018 and 2019, the organization held two International Conferences on Climate Change and Energy in Germany, with sponsors, speakers, and members all well associated with the countermovement including James Taylor of the Heartland Institute, Benny Peiser of the GWPF and Myron Ebell (CEI) (EIKE, 2019). President of the Institute Dr. Holger Thuss had previously reflected on the 2011 Heartland Institute Conference on Climate Change (No Trick Zone, 2011). Ultimately adopting the conference practice in Europe certainly mirrored its Heartland ally. Moreover, this program of events, with similarities to the USA, sees a more overt form of climate skepticism, although largely marginalized in its aim to impact policy legislation, whereby "solutions" rather than "denialism" have previously been more readily apparent in the narratives of CCCM in Europe. With that said, much like the GWPF in the UK, such fringe-like skeptic groups could potentially impact into the political realm. For instance, the best indication of EIKE's reach into the political sphere is the messages of EIKE echoed by the far-right political party Alternative for Germany (AfD) whereby there were profound similarities between AFD and EIKE outputs on the issue of climate change (Neujeffski, 2019). From 2009 to 2019, similar negative framing targeted environmentalists and environmentalism. Moreover, the report identified the frequent positive feedback towards the conservative alternative Greta Thurnberg, Naomi Seibt as a symbol of countermovement opposition, focusing on targeting young activists and recruitment via social media. Of specific note is the evidence of utilizing online platforms in transforming the opportunities for countermovements to expand their influence on the public. Thus, extending beyond traditional media outlets.

The far-right connections between climate skepticism in Germany reflect similar ties to those in other European countries. While a complete debate on the relationship between right-wing populism and climate change obstruction cannot be explored here, there is an alignment between Far-Right political parties and individuals with these countermovement positions that swept across Europe (for an overview, see Georgiadou et al., 2018). Moreover, as Ekberg et al. (2022) explore, far-right actors and conservative political parties have engaged in organized obstruction activities from the late 1980s to 2009. Of relevance here in the context of the CCCM are the noted funding ties to fossil industries and the role of the front group Climate Realists (previously the Stockholm Initiative) in Sweden. Climate Realist (Klimatsans) claimed that:

> *Unfortunately, we have to state that most news about the climate in the media is incomplete, misleading or completely wrong. The completely fundamental question of WHETHER the emissions of carbon dioxide affect the climate so much that it is noticeable is rarely addressed. It is presented instead as an established truth.*

Like the AFD, the far-right party in Sweden (Sverigedemokraterna) deployed discourses of delay in the political sphere in 2013 (see also Hultman et al., 2019).

The organizations in the ECRN, then, could best be described as fringe groups. Nevertheless, their messages and connections with contrarians from think tanks such as the Heartland institute would offer an appearance of legitimacy. For instance, ECRN member of the Irish Climate Science Forum (ICSF), founded in 2016, held a series of launch events, including contrarians affiliated with other CCCM Richard Lindzen (see Chapter 2) and William Happer. Emeritus Professor of Physics at Princeton, Happer was one of the founders of the USA-based CO_2 Coalition launched alongside Rodney W. Nichols, and Roger Cohen another fringe-like group in the USA. Moreover, created by the Dutch organization CLINTEL, the World Climate Declaration included signatures from many individuals found working for or promoted by other CCCM organizations including Richard Lindzen (Chapter 2), Christopher Monckton, and Ian Plimer (Clintel, n.d.). Although, the political impact of CLINTAL and ICSF is somewhat limited and unknown compared to the work of EIKE.

3.9 Conclusion

This chapter has explored the different countermovement organizations operating across Europe, mainly focusing on climate change from the 2000s onwards. Importantly, there is a precise series of transnational ties within Europe and how these organizations could expand their messages outside of the continent. Singular voices primarily diffused this, commentators or political educators on specific issues that gain access to the political realm and the media echo chamber to influence public perceptions and policies on climate change across the globe. While there was a significant concentration of countermovement organizations operating within Europe, their reach extends beyond European borders and was influenced by external organizations. Interlocking relationships have manifested, with ideas diffused across the globe. While we think of the USA CCCM as unique and organized, the reach of organizations based in Europe demonstrates the international capacity of the countermovement to extend its messaging across borders. This is arguably in alignment with the political thoughts of think tank organizations and individual actors, particularly economic academics that can dispense their discourses of delay.

References

Adam Smith Institute. (2009, May 29). *Professor Ian Plimer: Global warming*. https://www.adamsmith.org/blog/energy-environment/professor-ian-plimer-global-warming?rq=climate%20change. Accessed 24 Oct 2022.

Almiron, N., & Xifra, J. (2021). *Repensar los think tanks: Expertos vs. impostores= Rethinking think tanks: Experts vs impostors* (No. BOOK-2021-021). Universidad de Zaragoza.

Almiron, N., Boykoff, M., & Narberhaus, M. (2020). Dominant counter-frames in influential climate contrarian European think tanks. *Climatic Change, 162*, 2003–2020. https://doi.org/10.1007/s10584-020-02820-4

Álvarez, G. C. (2009, March 1). *Study of the effects on employment of public aid to renewable energy sources*. https://www.heartland.org/publications-resources/publications/study-of-the-effects-on-employment-of-public-aid-to-renewable-energy-sources?source=policybot. Accessed 31 Oct 2022

Anshelm, J., & Hultman, M. (2014). A green fatwā? Climate change as a threat to the masculinity of industrial modernity. *NORMA, 9*(2), 84–96. https://doi.org/10.1080/18902138.2014.908627

Austrian Economics Center. (n.d.). *Organization*. https://www.austriancenter.com/organization/. Accessed 11 Oct 2022.

Balling, R. C. (2002). *The global warming scapegoat*. https://web.archive.org/web/20030202172149/http://www.policynetwork.net/article.php?ID=404. Accessed 21 Mar 2022.

Bates, R. (1996). *The political economy of climate change science: A discernible human influence on climate documents*. Institute of Economic Affairs.

Bates, R., & Morris, J. (1994). *Global warming: Apocalypse or hot air? (Studies on the Environment). No. 1*. Institute of Economic Affairs.

Bessner, D. (2014). Murray Rothbard, political strategy, and the making of modern libertarianism. *Intellectual History Review, 24*(4), 441–456. https://doi.org/10.1080/17496977.2014.970371

Beunder, A. (2020, May 14). *De Smog (and) how shell backed Dutch 'Coordinator' of climate science denial for decades*. https://www.desmog.com/2020/05/14/bottcher-shell-funding-european-climate-science-denial/. Accessed 21 Mar 2022.

Böttcher, F. (1996). Climate change: Forcing a treaty. *Energy and Environment, 7*(4) https://doi.org/10.1177/0958305X9600700407

Bradley, R. L. (2003). *Climate alarmism reconsidered*. Institute of Economic Affairs.

Bowman, T. (2008, June 11). *Climate change and central planning*. https://www.adamsmith.org/blog/energy-environment/climate-change-and-central-planning?rq=climate%20change. Accessed 21 Mar 2022.

Bruno Leoni Institute. (n.d.). *Who we are*. https://www.brunoleoni.it/chi-siamo. Accessed 21 Mar 2022.

Carbon Brief. (2012). *Why would the Godfather of Tory donations fund the global warming policy foundation?* https://www.carbonbrief.org/why-would-the-godfather-of-tory-donations-fund-the-global-warming-policy-foundation. Accessed 21 Mar 2022.

Carter, N., & Clements, B. (2015). From 'greenest government ever' to 'get rid of all the green crap': David Cameron, the conservatives and the environment. *British Politics, 10*, 204–225. https://doi.org/10.1057/bp.2015.16

Carter, L., & Ross, A. (2018). *Revealed: BP and gambling interests fund secretive free market think tank.* https://unearthed.greenpeace.org/2018/07/30/bp-funding-institute-of-economic-affairs-gambling/. Accessed 24 Oct 2022.

Center for New Europe. (n.d.). *Who we are.* https://web.archive.org/web/20040725042502/http://www.cne.org/about2.htm. Accessed 21 Mar 2022.

Center for Political Studies. (2009, September). *Wind energy—The case of Denmark.* http://www.templar.co.uk/downloads/Wind_energy_-_the_case_of_Denmark.pdf. Accessed 21 Mar 2022.

CHEMEurope. (n.d). *Heidelberg appeal.* https://www.chemeurope.com/en/encyclopedia/Heidelberg_Appeal.html#_note-1/. Accessed 21 Mar 2022.

Clements, B. (2012). Exploring public opinion on the issue of climate change in Britain. *British Politics, 7*, 183–202. https://doi.org/10.1057/bp.2012.1

Clexit. (2016). *Committee members.* http://clexit.net/wp-content/uploads/2016/07/clexit-members.pdf

Clexit. (n.d.). *About.* https://clexit.net/about/. Accessed 21 Mar 2022.

CLINTEL. (n.d.). *About us.* https://clintel.org/about-us/. Accessed 21 Mar 2022.

Climate, Energy and Environment. (n.d.). *About us.* https://www.climato-realistes.fr/qui-sommes-nous/. Accessed 21 Mar 2022

Clintal. (n.d.). *World climate declaration.* https://clintel.org/wp-content/uploads/2022/08/WCD-version-0831224413.pdf. Accessed 21 Mar 2022.

Climate Realism. (n.d.). *Manifesto.* https://klimarealisme.dk/klimarealismens-manifest/. Accessed 21 Mar 2022.

Corporate Europe Observatory. (2010). *Secret funding for climate deniers.* https://corporateeurope.org/en/pressreleases/secret-funding-climate-deniers. Accessed 21 Mar 2022.

Culpepper, P. (2016). Capitalism, institutions, and power in the study of business. In O. Fioretos, T. Falleti, & A. Scheingate (Eds.), *The Oxford handbook of historical institutionalism* (pp. 453–466). Oxford University Press.

De Smog. (n.d.). *Institute of Economic Affairs.* https://www.desmog.com/institute-economic-affairs/. Accessed 21 Mar 2022.

De Smog. (n.d.). *International policy network.* https://www.desmog.com/international-policy-network/. Accessed 21 Mar 2022.

El Pais. (2008, September 17). *FAES edita un libro que cuestiona el cambio climático.* https://elpais.com/sociedad/2008/09/17/actualidad/1221602406_850215.html. Accessed 12 Dec 2022.

Ekberg, K., & Pressfeldt, V. (2022). A road to denial: Climate change and neoliberal thought in Sweden, 1988–2000. *Contemporary European History, 31*(4), 627–644.

Ekberg, K., Forchtner, B., Hultman, M., & Jylhä, K. M. (2022). *Climate obstruction: How denial, delay and inaction are heating the planet*. Taylor & Francis.

European Institute for Climate and Energy. (2007). *Die mission*. https://eike-klima-energie.eu/die-mission/grundsatzpapier-klima/. Accessed 21 Mar 2022.

European Institute for Climate and Energy. (2019). *Programm – 13. Internationale Klima- und Energiekonferenz*. https://eike-klima-energie.eu/programm/. Accessed 21 Mar 2022.

Fondaziane Magna Carta. (2010a). *"Global prosperity", l'altra faccia del cambiamento climatico*. http://magna-carta.it/articolo/global-prosperity-laltra-faccia-del-cambiamento-climatico/. Accessed 31 Oct 2022.

Fondaziane Magna Carta. (2010b). *A Copenaghen anche i leader occidentali invocano la catastrophe*. http://magna-carta.it/articolo/a-copenaghen-anche-i-leader-occidentali-invocano-la-catastrofe/. Accessed 31 Oct 2022.

Fondaziane Magna Carta. (2010c). *Secrets and Lies of "global warming."* http://magna-carta.it/articolo/segreti-e-bugie-del-global-warming/. Accessed 11 Oct 2022.

Frost, G. (2002). *Anthony Fisher: Champion of liberty*. Books Ltd.

Georgiadou, V., Rori, L., & Roumanias, C. (2018). Mapping the European far right in the 21st century: A meso-level analysis. *Electoral Studies, 54*, 103–115.

Global Warming Policy Foundation. (2009). *Mission statement*. https://www.thegwpf.org/who-we-are/. Accessed 21 Mar 2022.

Global Warming Policy Foundation. (2020). *Homepage*. https://www.netzerowatch.com/. Accessed 21 Mar 2022.

Goklany, I. (2005). *Clima: stabilizzazione o adattamento?* (Occasional Paper). Bruno Leoni Institute. https://www.brunoleoni.it/op-17-clima-stabilizzazione-o-adattamento

Hanley, S. (1999). The new right in the new Europe? Unravelling the ideology of 'Czech Thatcherism.' *Journal of Political Ideologies, 4*(2), 163–189.

Harris, M. (1996). The centre for policy studies: The paradoxes of power. *Contemporary British History, 10*(2), 51–64. https://doi.org/10.1080/13619469608581386

Heartland Institute. (n.d-a). *Christopher Monckton of Brenchley*. https://www.heartland.org/about-us/who-we-are/lord-christopher-monckton. Accessed 21 Mar 2022.

Heartland Institute. (n.d-b). *Holger J. Thuss*. https://www.heartland.org/about-us/who-we-are/holger-j-thuss. Accessed 21 Mar 2022.

Heartland Institute. (n.d-c). *Klaus, Vaclav*. https://www.heartland.org/about-us/who-we-are/vclav-klaus. Accessed 21 Mar 2022.

House Select Committee. (2009, September 24). Testimony of Gabriel Calzada Alvarez. Testimony before the House Select Committee on Energy Independence and Global Warming. Washington, DC. https://s.libertaddigital.com/doc/intervencion-de-calzada-en-el-congreso-de-eeuu-20210147.pdf. Accessed 21 Mar 2022.

Hüzeir, V., & Fraser, G. K. (2017). *A pipeline of ideas: How the Rotterdam School of Management facilitates climate change by collaborating with the fossil fuel industry*. Changerism.

Hultman, M., Björk, A., & Viinikka, T (2019). Denouncing environmental challenges via anti-establishment rhetoric, marketing of doubts, industrial/breadwinner masculinities enactments and ethno-nationalism. In M. Hultman (Ed.), *The far right and climate change denial*. Routledge.

Influence Map. (n.d.). *Shell*. https://lobbymap.org/company/Royal-Dutch-Shell. Accessed 13 Sept 2022.

Irish Climate Science Forum. (n.d.). *Homepage*. https://www.icsf.ie/. Accessed 18 Sept 2022.

James, S. (1993). The idea brokers: The impact of think tanks on British government. *Public Administration, 71*, 491–506. https://doi.org/10.1111/j.1467-9299.1993.tb00988.x

Johnson, B. (2000, November 2). Keep on truckin', folks—You've already won. *The Telegraph*. https://www.telegraph.co.uk/comment/4256411/Keep-on-truckin-folks-youve-already-won.html. Accessed 18 Sept 2022.

Johnson, B. (2013, January 20). It's snowing, and it really feels like the start of a mini ice age. *The Telegraph*. https://www.telegraph.co.uk/comment/columnists/borisjohnson/9814618/Its-snowing-and-it-really-feels-like-the-start-of-a-mini-ice-age.html. Accessed 18 Sept 2022.

Klaus, V. (2007, June 13). Freedom, not climate, is at risk. *Financial Times*. https://www.ft.com/content/9deb730a-19ca-11dc-99c5-000b5df10621. Accessed 18 Sept 2022.

Klaus, V. (2010). *The climate change doctrine is part of environmentalism, not of science*. https://www.klaus.cz/clanky/2694. Accessed 18 Sept 2022.

Klimarealistane. (n.d.). *Homepage*. https://www.klimarealistene.com/om-oss/. Accessed 18 Sept 2022.

Klimatsans. (n.d.). *Homepage*. https://klimatsans.com/. Accessed 18 Sept 2022.

Lawson, N. (2006a). *L'economia e la politica del cambiamento climatico Un appello alla ragione* (Occasional Paper). Bruno Leoni Institute. https://www.brunoleoni.it/op-36-l-economia-e-la-politica-del-cambiamento-climatico-un-appello-alla-ragione. Accessed 18 Sept 2022.

Lawson, N. (2006b). *Clima di superstizione* (Occasional Paper). Bruno Leoni Institute. https://www.brunoleoni.it/op-30-clima-di-superstizione. Accessed 18 Sept 2022.

Lilley, P. (2012). *What is wrong with stern? The failings of the stern review of the economics of climate change for the organization* (GWPF Report 9). Global Warming Policy Foundation. ISBN: 978-0-9566875-9-3.

Livermore, M. (2007). *Climate change: A guide to the scientific uncertainties.* Centre for Policy Studies.

McKie, R. E. (2018). *Rebranding the climate change counter movement through a criminological and political economic lens* (Doctoral dissertation, Northumbria University).

Meyer, D. S. (2004). Protests and political opportunities. *Annual Review of Sociology, 30*, 125–145.

Milkulski, B. A. (2011). *Hearing of the Committee of Health, Education, Labor and Pensions to the US Senate, on the issue of labour, working families and the 'American' Dream.* https://www.govinfo.gov/content/pkg/CHRG-112shrg86919/html/CHRG-112shrg86919.htm

Morris, J. (2005). *Memorandum by professor Julian Morris, University of Buckingham [84].* https://publications.parliament.uk/pa/ld200506/ldselect/ldeconaf/12/12we19.htm. Accessed 13 Dec 2022.

Morris, J. (2006). *The role of market institutions in enabling adaptation to climate change.* https://reason.org/reason-roundtable/the-role-of-market-institution/. Accessed 21 Mar 2022.

Morris, J. (2013). *Which policy to address climate change.* International Policy Press.

Neujeffski, M. (2019, May 15). *Dubious experts: EIKE's climate change deniers and the AfD Climate Symposium.* http://thinktanknetworkresearch.net/blog_ttni_en/dubious-experts-eikes-climate-change-deniers-and-the-afd-climate-symposium/. Accessed 21 Mar 2022.

No Trick Zone. (2011). *Dr. Holger Thuss of EIKE reports on 6th International Climate Change Conference in Washington.* https://notrickszone.com/2011/07/08/dr-holger-thuss-of-eike-reports-on-the-6th-international-climate-change-conference-in-washington/. Accessed 21 Mar 2022.

North, R. E. (2010). *Speaking up for BP: Cross your fingers but be patient.* https://iea.org.uk/blog/speaking-for-bp-cross-your-fingers-be-patient. Accessed 21 Mar 2022.

Painter, J., & Ashe, T. (2012). Cross-national comparison of the presence of climate skepticism in the print media in six countries, 2007–2010. *Environmental Research Letters, 7*(4). https://doi.org/10.1088/1748-9326/7/4/044005

Patty, B. (2010, December 4). *Sense needed in climate change talks.* https://www.adamsmith.org/blog/energy-environment/sense-needed-in-climate-change-talks?rq=climate%20change

Plehwe, D. (2021). Think tanks and the politics of climate change. In D. E. Abelson & C. J. Rastrick (Eds.), *Handbook on think tanks in public policy* (pp. 150–165). Edward Elgar Publishing.

Plehwe, D., Goldenbaum, M., Ramanujam, A., McKie, R. E., Moreno, J., Ekberg, K., Hall, G., Araldi, L., Walker, J., Brulle, R., Neujeffski, M., Graham, N., & Hrubes, M. (2021, July). *The mises network and climate policy* (Policy Briefing). The Climate Social Science Network.

Pons-Hernández, M. (2022). Power (ful) Connections: Exploring the revolving doors phenomenon as a form of state-corporate crime. *Critical Criminology, 30*, 1–16.

Sourcewatch. (n.d.). *European academy for environmental affairs*. https://www.sourcewatch.org/index.php/European_Academy_for_Environmental_Affairs. Accessed 6 Dec 2022.

Sowel, N. (2007). The economic policy agenda of the Conservative party. *The Observatory of British Society, 4*, 131–153. https://doi.org/10.4000/osb.374

Singer, F. (2009). *Climategate: Some comments*. https://iea.org.uk/blog/climategate-some-comments-by-prof-s-fred-singers. Denouncing environmental challenges via anti-establishment rhetoric, marketing of doubts, industrial/breadwinner masculinities enactments and ethno-nationalism.

Science, Climate and Energy. (n.d.). *Home*. https://www.science-climat-energie.be/home/. Accessed 21 Mar 2022.

Source Watch. (n.d.). *Center for the New Europe*. https://www.sourcewatch.org/index.php?title=Centre_for_the_New_Europe. Accessed 21 Mar 2022.

Smaling, E. (2019, February 27). RSM ends controversial contract with Shell. *Erasmus Magazine*. https://www.erasmusmagazine.nl/en/2019/02/27/rsm-ends-controversial-contract-with-shell/. Accessed 21 Apr 2022.

Ströher, B., & Peiser, B. (2005). Report on the climate change seminar of the Friedrich Naumann Foundation, Gummersbach, Germany. *Energy & Environment, 16*(3–4), 681–683. https://doi.org/10.1260/0958305054672268

The Mises Institute. (n.d.). *Gabriel Calzada Alvarez*. https://mises.org/profile/gabriel-calzada-alvarez. Accessed 21 Mar 2022.

Tobacco Tactics. (2020a). *Institute for Economic Affairs*. https://tobaccotactics.org/wiki/institute-of-economic-affairs/. Accessed 3 Mar 2022.

Tobacco Tactics. (2020b). *European science and environment forum*. https://tobaccotactics.org/wiki/european-science-and-environment-forum/. Accessed 3 Mar 2022.

Tupy, M. L. (2017, October 31). *How the debate on climate change is cooling down*. https://www.austriancenter.com/debate-climate-change-cooling/. Accessed 3 Mar 2022.

CHAPTER 4

Australia and New Zealand: Fossil Fuels and Climate Mitigation Failings

4.1 Introduction

This chapter examines the contours of CCCM organizations in Australia and New Zealand. First, tracing the history of countermovement opposition in Australia, the chapter highlights that the country lagged in its commitments and implementing mechanisms to reduce GHG emissions. As a result, CCCM organizations would play a role in disseminating discourses of delay supported by the activities of specific individuals connected directly with fossil fuel companies that would impact climate and environmental legislation since the Howard government in 1995. These initial insight illustrates how the Howard Government was on the side of the fossil fuel industry before Kevin Rudd and the left-of-center Labor Party attempted to introduce key climate-related legislation. Notably, the role of the Murdoch press took center stage during Rudd's tenure to reinforce fossil capitalism's role in the political economy of Australia. However, Rudd's actions were backtracked by the rise of Tony Abbott and during his tenure as prime minister, where traditional forms of climate change denial would flourish used by himself and Australian countermovement organizations. Following, it explores the several overlaps with New Zealand countermovement opposition, distinguishing how similar fringe campaigns rallied the public and political thought focused more specifically on the agribusiness sector rather than the fossil fuel industry. Therefore, while central to the chapter and book is the role

R. E. McKie, *The Climate Change Counter Movement*,
https://doi.org/10.1007/978-3-031-33592-1_4

of fossil fuels, it illustrates the evolving complexity of countermovement operations that recognized the political-economic context of New Zealand and how a similar operation related to a different business sector, yet one concerned with the issue of climate, could successfully undermine climate action.

4.2 Coal and Gas Mining

While Australian public opinion on the issue of climate change had consistently been one of concern (e.g., Leviston et al., 2011), the country lagged in its commitments and implementation of mechanisms to reduce GHG emissions (Climate Action Tracker, n.d.). One reason is that coal and gas mining had been crafted in the political mind as a resource central to the Australian political economy. As a resource-intensive nation, coal was framed and aligned with Australia's social and cultural makeup (Curran, 2021). As Wright et al. (2022) outline, the fossil fuel industry used "articulatory practices, [where] fossil energy is constructed as fundamental to economic growth and affluence, despite the obvious scientific discourse that fossil fuels are the dominant producer of greenhouse gas emissions" (p. 547). In turn, fossil fuels would be framed as central for community and collective wellbeing, and any attempts at decarbonization would be an economic catastrophe for the future of Australia. Therefore, the issue of addressing climate change became part of the conversation where energy and climate-related policy aligned with how its impacts would be perceived to influence everyday lives.

When the UNFCCC 1992 conference event occurred, Australia responded with its first written report. The report indicated that Australia's future response to climate change would be entangled with its history as primarily a fossil fuel economy. This is because, they directly questioned whether they should be included in Annex I, like most developed countries (Community Information Unit of the Department of the Environment Sport and Territories, 1994). With the issue of climate change on the table for Australian policymakers, before and after the Howard government's election in 1995, it was clear the impact Australia's fossil fuel-based political economy would have on their response to climate change. As Lawrence (2009) notes, under the Howard Government (1995–2007), "Australia was largely dictated by fossil fuel and mineral sector interests and reacted a close alliance with the Bush administration" (p. 281). Thus, the Howard government, in contrast to the previous Labor government, aligned its priorities with the Bush administration and failed to ratify the 1997 Kyoto Agreement alongside acquiring "a generous target in the Kyoto Protocol negotiations" (p. 283). Deploying discourses of delay directly into policy proposals, the Howard Government claimed that ratifying Kyoto would burden the Australian economy and have limited effect if countries such as India and China did not have such binding targets. Ultimately, lobbying by energy-intensive industries (Pearse, 2007) and framing mining coal and gas as a defining feature of the political economy of Australia (Christoff, 2022) stalled and stunted these early opportunities to address rising GHG emissions (see also Lucas, 2022).

Christoff (2022) outlines five groups of actors that helped shape climate and energy policies in Australia. These included political actors (politicians and political parties), knowledge brokers (climate scientists, economists, and other academics), national environmental NGOs, the mass media, and energy industry representatives, and lobbyists. Each of these groups makes up parts of the Australian arm of the CCCM that would, over time, be able to influence politicians, policy decisions, and public opinions on climate change.

4.3 The Howard Government: The Power of the Extractive Industry (1995–2007)

Given the increasing partisanship in Australian politics on the issue of climate change, it was unsurprising when coupled with effective lobbying (Lucas, 2022), the fossil fuel industry operations and supporting network of actors would have some success in delaying climate action. Lucas (2021) identified top organizations where current or former politicians, staffers, or senior bureaucrats occupied the industry. These organizations included The Minerals Council of Australia, QLD Mining Corporation, the Association of Mining and Exploration, and the Australian Coal Association. These organizations had played a historical role in shaping policy and interplaying between government and industry. For example, Hamilton (2009) describes the significant intimacy between the fossil fuel lobby and the Howard Government and the revolving door of government individuals, fossil fuel companies, and related CCCM organizations. Also, Ralph Hillman, formerly Ambassador for the Environment under the Howard Government, became the Executive Director of the Australian Coal Association (2007–2011) (see Global Energy Monitor

Wiki, n.d.; Milman, 2011). In 2001 in his role as Ambassador for the Environment, he gave a speech to the Australian Institute of International Affairs claiming in response to the limited constraints put towards development and limiting GHG emissions in developing countries, Hillman claimed:

> *An international response to climate change will not survive for long if some countries have a price on carbon and others do not...Australia would lose competitiveness and new investments would move to countries not bound by emissions constraints...cuts in Australia were offset by increased emissions from these relocated industries.* (2001, n.p.)

While proposing to implement policies that would address the climate crisis, this speech reflected some of the delaying discourses to challenge ratifying Kyoto, especially given at the time, the Bush administration's withdrawal would stall global efforts to address climate change.

Moreover, Pearse (2011a) outlined the influence of the *Greenhouse Mafia* and the role played by the lobbying group Australian Industry Greenhouse Network (AIGN). The AIGN included the Australian Coal Association, Australian Petroleum Production and Exploration Association, Federal Chamber of Automotive Industries, and Australian Institute of Petroleum (ChemEurope, n.d.). Much like the industry representatives for fossil fuel corporations in the USA, these associations were able to penetrate the Howard Government's approach to climate and energy-related policies. This is best evidenced in response to the Kyoto Protocol. As Christoff (2005) recalls in response to the Howard government's White Paper: *Securing Australia's Energy Future,* was: "*an act of policy schizophrenia by acknowledging and benchmarking itself against the Kyoto Protocol,*" *which had been signed* "*but also repeatedly opposed, undermined and refused to ratify*" (p. 29).

In 2002, the Australian Coal Association (ACA) developed the narrative that clean coal could solve climate change. As a solution, it drives economic growth through investment in technological innovations and avoids the fuel phase-out. Citing Nigel Wilson's *Clean Coal may be the fuel of the future* from the Weekend Australian (2003)—the only national news outlet owned by the Rupert Murdoch Press (see below)—the ACA conveyed that they were: "*moving to ensure its future is not destroyed by increasing concerns about GHG emissions and the wide application of the Kyoto Protocol*" (n.p.). At the time, the paper identified that the ACA was working with the Australian government with reported outcomes, including backing the COAL21 agenda to develop technological innovations in the coal industry by 2021 from then Federal Industry Minister Ian MacFarlane. Further media releases highlighted by the association continued to promote the "cleaning of coal" and part of the solution (see ACA, 2003), often promoted by Mark O'Neill, previous director of the ACA (e.g., ABC, 2007).

In addition, by playing a role in advancing countermovement narratives, the Center for Independent Studies (CIS) began to engage in a campaign that would help support and undermine climate legislation. The CIS was founded in 1976, "*influenced by libertarian thinkers such as Murray Rothbard, Friedrich Hayek and Milton Friedman as well as classical liberal philosophers such as Adam Smith, David Hume and John Locke...*" (n.p.). Greg Lindsey established the think tank as an Australian version of Anthony Fishers IEA in the UK. The organization held close ties with the Howard Government, and even on its 20th anniversary in 1996, it held a special event, including then prime minister Howard. Howard said, "*it has made better policy...We have better governments on both sides of the political equation as a result...*" (CIS, 1996). The CIS produced its journal *Policy,* which incorporated papers by its fellows and others. Importantly, in 1997 they extensively discussed economics and the environment, including aspects focusing on climate change. As such, the CIS framed climate change as primarily an economic issue. They also criticized the media and environmental activists and promoted the work of others who sometimes critiqued the scientific evidence on climate change. For instance, in the lead-up to the Kyoto Protocol, the CIS promoted, in their Policy journal, a piece by Geoff Hogbin, an economist by trade:

> *Rhetoric leading up to the Kyoto Conference, have aroused widespread fears of global warming as a consequence of the build-up of greenhouse gases, primarily* CO_2, *caused by capitalism's voracious consumption of fossil fuel energy. Exploiting these anxieties, lobby groups and the media have put governments under strong pressures to cut consumption of fossil fuels with the objective of curbing greenhouse gas emissions.* (1998, n.p.)

Following this, the CIS continued to present commentary from authors such as Lomborg (Lomborg, 2006) and other CIS fellows (e.g., Maley, 2000). Such reports included critiques of the science while over-representing the implications of Australian climate and energy policy. For example, in 2007, the center produced a report by John Humphreys describing the impact of the proposed Australian Carbon Tax. Humphreys was an economist and politician for the Liberal Democratic party, and his ideological positions fed into the text claiming,

> *The most efficient way to do this is to introduce a price signal and allow the market to determine the best alternative. The government should not attempt to pick winners or to bias the market in favor of any alternative such as nuclear, wind, solar, 'clean' coal, or hot rocks, and funding for these industries should be removed. A price signal can be introduced either through a carbon tax or through carbon trading.* (Humphreys, 2007, p. 7)

Thus, climate change solutions centered on market forces over government intervention aligned with the CIS' ideological positions. Although it was not simply the ideological position that clarified the CIS would be deployed as a CCCM. As Walker (2021) outlined the funding to establish the CIS came from fossil fuel and mining capital, including donors "Hugh Morgan (Western Mining Corporation), John MaCloed (Conzinc-Rio Tinto) and Douglas Hocking (Shell), John Brunner.

(BHP), and John Bonython of Murdoch's Adelaide Advertiser Group, cofounder of gas giant Santos" (p. 205). The evident combination of fossil fuel interests and neoliberal thought ideas suggests a clear lineage between fossil fuel extraction and the advancement of the neoliberal regime in Australia that would ultimately delay action to address the climate crisis.

4.4 Combatting Kevin Rudd's Government: The Institute for Public Affairs

Following the Howard government's loss at the election, the Australian Labor Party's Kevin Rudd became Prime Minister from 2007 to 2010 and later from 2013 to 2015. The more center-left party made some progress on climate and energy policy, ratifying the Kyoto Protocol and attempting to introduce the Carbon Pollution Reduction Scheme. Nevertheless, despite an optimistic outlook, there was pushback and resistance. Ultimately, the cabinet's legislative and progressive agenda sparked heightened engagement from the CCCM.

One example of an organization that campaigned aggressively against the Rudd government's climate proposals and policy was the Institute for Public Affairs (IPA). The IPA, founded in 1943, promoted a "*free market of ideas, the free flow of capital, a limited and efficient government, evidence-based public policy, the rule of law, and representative democracy.*" (IPA, n.d., n.p.). In response to the growing power of the Labor Party in post-WWII Australia (IPA, 1968), they highlighted, "*It soon became apparent that the IPAs efforts could best be exerted outside the field of direct political participation*" (p. 3). Thus, the think tank took on a role, much like the IEA, to disseminate "ideas" not necessarily through the traditional political system yet still into politics.

Several businessmen linked with the fossil fuel industry-funded the IPA. These included Charles Goode (Chairman of Woodside Petroleum 1997–2007), William Morgan (former CEO of Western Mining), and Harold Clough (former manager and director of Clough Ltd), to name a few (Beder, n.d.). Rupert Murdoch's father, Keith Murdoch, was also a founder, and ultimately the IPA can best be understood as a source for mining propaganda like the CIS (Walker, 2021). At the same time, it was an essential component in elevating the Liberal Party. Like the CIS, the IPA promoted neoliberal ideas in Australia that would impact and shape policy (Mendes, 2003). The IPA merged with the libertarian think tank the Australian Institute for Public Policy due to its founder John Hyde's election. What remained was a conservative, libertarian corporate-funded think tank that would go on to disseminate reports and engage in campaign and media activities to undermine climate action.

Furthermore, the positions and connections between the IPA and the fossil industry became even more prominent when Gina Rinehart, director of the mining company Hancock Prospecting, donated $4.5 million to the IPA between 2016 and 2017 (Readfearn, 2018). However, we can speculate that Rinehart was also an undisclosed donor to the IPA before this finding (Walker, 2021). The IPA had played a role in the Tobacco Industry Lobby (Tobacco Tactics, 2020) and it was unsurprising that it would become part of the wider CCCM.

Under Prime Minister Gillard (Labor) (2010–2013), the administration aimed to introduce the Clean Energy Future carbon pricing scheme that would cut Australia's emissions significantly. As Pearse (2011b, n.p.) outlines, however:

> *"Money explains the behavior of many campaigning against Gillard, as those in her corner are quick to highlight; the proudly skeptical and coal-friendly Institute of Public Affairs, for example, has admitted they rarely take a position different from the dozen energy firms who contribute funds to them, because otherwise they'd stop funding us."*

Although Pearse emphasized the conflicts within pro-environmental movements in response to the government's plans, the IPA and related fossil fuel industry actors would aim to present a critique of the proposed mechanism and indirectly influence the political process. The IPA submitted feedback regarding the scheme (Moran, 2010) and published press releases on the issue (e.g., IPA, 2012). Moran was the author of the IPA's edited collection *Climate Change: The Facts* and the text included essays by familiar individuals that were promoted by or worked for other organizations involved in the countermovement, including Willie Soon, David Legates, and Nigel Lawson. This ensured that as a think tank, the IPA was disseminating policy-focused arguments which challenged climate action and environmental legislation. However, texts such as *Climate Change: The Facts* deployed science-based skepticism often reserved for advocacy and fringe-like countermovement organizations that would most often deny or minimize the impact of CO^2 emissions on the climate.

Central to developing its outreach and mainstreaming the focus on climate change, the IPA created the Australian Environment Foundation (AEF) front group. The organization claimed to: "*take an evidence-based, solution-focused approach to environmental issues*" that was "*caring for both Australia & Australians*" (n.d., n.p.) Like the Heartland Institute, the AEF held yearly conferences from 2006 (AEF, 2006). Speeches at their first conference had little discussion on climate change. Nevertheless, the then Senior Fellow of the IPA, Jennifer Marohasy, presented a paper explaining the critical role of the AEF in disseminating oppositional positions:

> *We all accept that carbon dioxide has the potential to warm the earth and like Al Gore, I am concerned about elevated levels of atmospheric carbon dioxide from the burning of fossil fuels. But even if Kyoto were fully implemented, with Australia and the USA signing up, there would be virtually no potential impact on climate because about 70 percent of carbon dioxide emissions are from countries not subject to Kyoto restrictions...Global warming is not a*

moral issue, but a technological issue...It's fair to conclude there will be a worldwide transition from oil to something else, but we don't know how rough or smooth this transition might be, nor whether in 20 years' time ethanol, or hydrogen, or something else, will be the dominant transport fuel. (pp. 2, 6–7)

The 2007 and 2008 events drew more attention to the issue of climate change with sessions led by Dr. Chris de Freitas (de Freitas, 2007), Bob Carter (2007), and William Kininmonth, who explained the launch of the Australian Climate Science Coalition (ACSC) (AEF, 2008) (See Below on the ACSC).

The AEF membership did slightly change over time. Still, key figures included Peter Ridd (previous science coordinator), Jennifer Marohasy (previous chairman, director) (AEF, 2005), and Tom Bobstock (previous director) (AEF, 2007) remained on board. Aforementioned, Marohasy was also a Senior Fellow at the IPA, and Bobstock and Ridd took roles in other CCCM organizations. For instance, Bobstock was a director of the Lavoisier Group (see below) (AEF, 2007) and Ridd, a Geophysicist, a member of the academic advisory panel for the CCCM organization the GWPF, an author in an edited volume from the IPA (Ridd, 2017) and advisor to the CCCM organization the Galileo Movement (see below). Thus, while the AEF claimed to be a conservation and environmental group, it was a CCCM organization that actively campaigned to undermine climate-related legislation.

Walker (2021) describes the origins of fossil neoliberalism in Australia. In particular, he draws attention to the Australian economist Torleiv Hytten, who was invited to the first meeting of the Mont Pelerin Society (MPS) (Chapter 6). Hytten himself worked closely with the IPA and noting Murdoch additionally would be a fellow of the MPS, it was unsurprising that the IPA became and outlet that spoke specifically for the fossil fuel industry in Australia. In fact, in 1976 Hayek (MPS fellow) would emphasize the use of counterscience rhetoric across issues. What followed would be the IPA's unsurprising attempts to present examples of scientific denial primarily through its front group the AEF as well as its specific policy positions oriented to make mainstream political impact.

4.5 Fringe Groups, Coal Links, and Public Engagement

Much like the advocacy groups that garnered mainstream attention, which spread misinformation and challenged the science of climate change in the USA and Europe, similar organizations operated in Australia during the Rudd and Gillard governments. For example, the Carbon Sense Coalition was founded in 2007 with Viv Forbes as executive director. Forbes, a former General Manager for the coal company Rochlands Richfield Ltd (Bloomberg, n.d.), had also taken on directorships and senior executive positions for other coal mining companies, including Smith Blackwater, and Austral Coal (Rocklands Richfield, 2006). This indicated the links between this advocacy group and the fossil industry.

Over time, the coalition promoted climate change skepticism, claiming positives to the increase in GHG emissions (e.g., Evans, 2002; McGauchie, 2000). They critiqued existing and proposed plans to curb GHG (Carbon Sense Coalition, n.d.-b) and promoted a reading list of work by individuals such as David Archibald, Christopher Monckton, and Fred Singer that had deployed discourse of delay (for this reading list, see Carbon Sense Coalition, n.d.-a). From 2007 until 2015, the coalition targeted different proposed climate strategies by the Australian Government to undermine the science. For example, in 2007, the coalition organization released *Look Before You Leap*, a response to the proposed Climate Smart 2050 Queensland Energy Policy (Carbon Sense Coalition, 2007a). They collated a series of points on the scientific debate claiming other factors had influenced the climate, and climate scientists were promoting "*prophecies of doom*" (Carbon Sense Coalition, 2007b, p. 3). They ended by suggesting that society will adapt by finding innovative solutions.

Similarly, in 2010, the organization released a report authored by Forbes that highlighted the positives and importance of coal (Carbon Sense Coalition, 2010a), and in 2015 released a newsletter arguing that the Paris Accords threatened Australia's access to energy, concluding:

> *"We must defend our right to develop and use our own land, resources, and property to produce abundant, reliable, and affordable energy and food…Tell our politicians: Sign NOTHING that we can't get out of in Paris in December."* (n.p.)

It may well be unsurprising then, that the coalition sought to influence legislation and challenge the proposals of the Rudd and Gillard government, given Forbes connections with coal companies (e.g., Stanmore Coal, n.d.). Therefore, Carbon Sense Coalition utilized the political opportunity of specific policy developments that emerged at one point. In essence, the pressure of environmental groups alongside national level policy that became more amenable to addressing the climate crisis would instigate an orchestrated campaign to undermine this particular policy proposal.

Additionally, Forbes was a scientific advisory board member of the ACSC. The ACSC claimed to encourage an "exchange of scientific ideas and to encourage proper political and social debate on this intriguing subject," reflecting that "past and current climates are sufficiently well understood to enable projections of future climate changes to be accurately predicted." (2009, n.p.). From its inception in 2008, much like other fringe CCCM groups, the organization promoted the work of several individuals that had disputed the scientific evidence on climate change or posed the question of a more timid response, such as the *Skeptics Handbook* written by Joanne Nova (2009). Nova also had a blog (https://joannenova.com.au/) that described the problems of climate science, promoting "apparent" evidence which challenged the positions of environmental groups and those advocating for appropriate policies to limit GHG emissions.

From 2008, the ACSC ensured its messages would reach the Australian media, from columns in the Australian (e.g., Nicol, 2008) and ABC National Counterpoints (e,g, Counterpoints, 2009). This media outreach was also essential for the ACSC, much like it was for the AEF to mainstream their opposition and find public support for these positions (for a list of media outputs, see AEF, n.d.). However, while the IPA would disseminate policy-related discourse rather than focus on climate skepticism and denial, the AEF and the ACSA would ensure a stream of information to the public that added the notions of skepticism and doubt in the science. Importantly, this was a starting point when the issue of climate change began to foster even greater political partisanship evident in the media (Fielding et al., 2012).

Archibald and Carter, on the scientific advisory board to ASCA, were connected to the Lavoisier Group (Lavoisier Group, 2000). The Lavoisier Group was founded in 2000 by Ray Evans, an executive at Western Mining Corp. Hugh Morgan, CEO of Western Mining Corp, gave the inaugural speech (Source Watch, n.d.). It could best be described as an advocacy organization that promoted messages that questioned climate science and challenged environmental regulations. Initially, they targeted policies on Kyoto targets (e.g., Evans, 2000). Then, in 2000, they submitted to response to the Senate References Committee for the Environment, Communications, Information Technology, and the Arts in response to the *Renewable Energy (Electricity) Bill Renewable Energy (Electricity) (Charge) Bill 2000*. In their response, they outlined that the Kyoto Protocol was a "bureaucratic folly" and "symbolic environmentalism that would appease some of the Government's critics" (n.p.). They similarly submitted a response to the JSCOT Inquiry into The Kyoto Protocol. In it, the organization invoked notions of sovereignty. The organization reinforced a sense of coal nationalism, "...the Lavoisier Group argues that with the Kyoto Protocol, we face the most serious challenge to our sovereignty since the Japanese Fleet entered the Coral Sea on 3rd May 1942." The emphasis in both submissions was a critique of international governance mechanisms and their impact on Australian economic policy. Moreover, it closely aligned with the positions and coal industry associations during the Howard government.

The focus on sovereignty served as an everyday discourse of delay that the organization would continue to deploy during Kevin Rudd's time in office. For instance, in 2007, the organization promoted the work of Ray Evans and his response to the IPCC and AR4, claiming:

> *Taxing the generation of coal-based electricity to render it so expensive as to seriously reduce its consumption will bring about huge economic dislocation... since no board of directors can sign off on an investment worth, say, $10 billions if a carbon tax as proposed by Kevin Rudd should suddenly render that investment uneconomic.* (Evans, 2007, p. 3)

During this period, the Rudd government focused on reshaping Australia's relationship with coal, attempting to introduce appropriate legislation that, at least on paper, would go towards meeting international targets. As a result, framing the actions of the Rudd government as a significant divergence from the success of Australia's fossil fuel political economy was unsurprising.

By ultimately connecting the economics of climate change, it aligned more closely with that of the Australian Liberal party and politicians. Even fringe-like groups, alongside think tanks, used these similar discourses of delay. An example is the Bert Kelly Research Center which managed the Lavoisier Group website. Charles Robert "Bert" Kelly was one of the first neoliberal MPs in Australia asserting the importance of free markets over the "socialist" policies of a Labor government (Kelly, 1978). The center promoted the book launch of contrarian Ian Plimer's: "How to get expelled from school—a guide to climate change for pupils, parents, and punters," (2011) jointly sponsored by the Bert Kelly Research Center, the Institute of Public Affairs and the National Civic Council" (Bert Kelly Research Center, 2011, n.p.). A forward of the text written by Vaclav Klaus (Chapter 3) aligned the skepticism to these economic ideas:

> *There are other mechanisms behind the global climate, and that the role of man-made Co2 emissions is-if any-very small, almost negligible. There is no doubt that the scientific debate about this issue is still very much ongoing but the public and political debate was prematurely closed and declared decided.* (p. 2)

The Bert Kelly Research Center could have been considered a legitimate policy/think tank force. However, it did have connections to fringe groups and individuals that could, in essence, further legitimize climate change denial. Thus, climate skepticism and forms of policy positions operationalized by the CCCM could infiltrate the political space using think tanks such as the IPA and Bert Kelly Research Center combined with front groups such as the AEF, ACSC, Carbon Sense Coalition, and the Lavoisier Group. However, it's important to note that at the time, this activity may have had little influence directly on policy.

4.6 The Abbott Government, Murdoch, and the Media

Fielding et al. (2012) emphasized the growing partisanship on the issue of climate change among Australian politicians from the early 2000s. The Labor Party considered on the left or center-left was, on the surface, more likely to support and advance effective climate policy than the right-of-center Liberal Party given that for success it required a coalition with Green party officials and voters. While this growing partisanship between politicians and political parties may have appeared, Crowley (2021, p. 4) emphasized that even though "*Labor is less beholden to climate skeptics and conservatism…it struggles to develop policy and experiences the same structural and political constraints as the Coalition.*" These structural constraints were related to Australia's identity as a coal nation and the fossil industry's hold over Australian policy (see Lucas, 2021). Nevertheless, while in the political realm, it appeared that there would be a consensus that restrictions on coal emissions could affect Australia's economic growth, the partisanship accelerated during the rise of Tony Abbott's government facilitated by an increasingly partisan media.

Anthony John "Tony" Abbott served as the 28th Prime Minister of Australia leading the Liberal Party (Parliament of Australia, n.d.), and his likely success was partly attributed to the media campaign, particularly the campaign's use of social media (Hurcombe, 2016). This was coupled with News Corp's all-out campaign against the Labor Party, that clearly benefited Abbott's campaign alongside funding from fossil fuel-related donors seeking to back and support a candidate that would in turn obstruct effective action on climate change (Mayne, 2015).

Taking this opportunity of growing partisanship, the Galileo Movement, founded in 2011, was an advocacy group that mirrored the fringe-like groups such as the GWPF and the Climate Realist Network. The organization claimed to: "expose misrepresentations of global warming pushing a price on carbon dioxide. We care about freedom, security, the environment, humanity and our future" (2012a, n.p.). On the UN Agenda 21, they claimed that:

> *UN Agenda 21 in particular, is a highly politicized umbrella campaign encompassing three parts: biodiversity, sustainability; fraudulent climate alarm as the crisis used to stir public support. All three parts rely on claims contradicting empirical scientific evidence. UN Agenda 21 is a politically driven campaign to undermine Australian sovereignty and governance.* (2012b, n.p.)

Members of the public Case Smit and John Smeed founded the organization based on the premise that they had "accepted claims of global warming due to human activity…[but] became aware of inconsistencies in the political claims and spin" (n.p.). Accordingly, the organization facilitated a campaign to disseminate discourses of delay and rally public members that could influence voter choices in an increasingly partisan political environment. The first initial drive to develop and solidify the organization as part of the countermovement came when the organization managed a tour with climate contrarian Christopher Monkton (Galileo

Movement, n.d.). A series of rallies followed, providing a voice for people who were "afraid to speak out publicly" (n.p.). Between 2011 and 2012, the organization held rallies nationwide, including an event in Sydney where Tony Abbot was in attendance (Galileo Movement, 2012c). Before even taking office then, Abbott engaged with this climate skeptic group, corresponding with his positions and policy approaches to addressing climate change that would later manifest in his administration.

The organization drew attention to specific policies towards the end of the Rudd and Gillard leadership. Before the 2011 election, they reported on a submission made by a member of the public to the local government (Gallileo Movement, 2011). The submission was on the proposed carbon tax to the Office of Senator Claire Moore (Labor) that requested the Senator engage with the work of the Galileo Movement. In response, Moore's office highlighted that the Galileo Movement was a climate skeptic organization. Following and during the 2012 election year, a political opportunity to enhance their message, they extended this critique of the Carbon Tax, claiming it would increase the costs of electricity, water, fuel, and grocery bills of the Australian population (Galileo Movement, 2012d). Thus, the arguments presented by the Galileo Movement were those echoing the political realm, media, and some public views on the issue.

Playing an essential role in Abbott's election was the well-constructed and expanding polarization in the media landscape, driven ultimately by the strength of Murdoch's and other right-wing news outlets (White, 2014). Explicitly linked with the coal industry and acting as a vehicle for mining propaganda, Murdoch could influence politics on climate change using his media empire (Walker, 2021). Young (2017) noted News Corp became unusually partisan in 2013, reflecting similar partisanship to the UK and Murdoch's Daily Mail and support for the Conservative Party. Young (2017) also emphasized that this became part of a broader shift for newspapers to adopt strong opinions. Therefore, News Corp heightened affiliation to the conservative side of Australian politics, demonstrating a more "*overt partisanship in reporting politics*" (p. 886). Likewise, White (2014) argued that, in part, because of its ability for Murdoch to influence the media outlets in Australia: "the country is now governed by a deeply unpopular Liberal-National government, crafted in the image of the most climate-denying elements of the Tea Party" (n.p.). What White argues was a Tea Party movement, also aligned with the activities of the

Tea Party in the USA as they continued to challenge Obama's legislative plans from the beginning of his tenure.

At the same time, media segments became the fossil fuel industry's PR campaign. Wright et al.'s (2021) analysis of newspaper articles in the Australian media between 2008 and 2019 illuminated this activity. Between 2008 and 2016, the fossil industry used the media to frame the industry as central to the national interest, reinforcing the Australian coal industry's historical, social, and cultural importance. One significant moment is Gina Rinehart and the battle over a carbon tax on Australian TV and Newspapers. Milman (2011) reports on the Bolt Report, conservative commentator and climate skeptic Andrew Bolt. Bolt signed with Ten Network to deliver a children's TV program, a news channel where Gina Rinehart invested in the network and became a board member.

Moreover, other news outlets facilitated opposition to the carbon tax, including Sydney's Sunday Telegraph and The Australian. This PR campaign would help diffuse delaying discourses that favored the interests of the fossil fuel corporations and interested politicians by appealing to "a common identity, in which resource extraction was presented as central to national identity and character" (Wright et al., 2021, p. 3). As such, Australian media ultimately helped facilitate the PR campaigns of the fossil fuel industry to help normalize and reinforce the hegemony of the coal industry and its role in Australia's political economy.

The impact of increasing media partisanship, lobbying activities associated with the fossil fuel industry, and the CCCM can best be observed during Abbott's tenure as Prime Minister. From his entry into power in 2013 to 2015, the Abbott government repealed the carbon price legislation (Foyster, 2019) and disbanded the Climate Commission. Yet despite this disruption and policy uncertainty introduced at the federal level. MacNeil (2021) noted that subnational governments nationwide would fill the policy vacuum in response to Abbott's dialing back of national climate strategies by continuing to support the massive boom in renewable energy investment triggered by the Rudd/Gillard administration. Nevertheless, it is fair to assume that at a federal level, Abbott's government was much like the Howard government, at the behest of the fossil fuel economy, fossil fuel corporations, and vested interests in protecting the hegemonic political economy despite the growing crisis of climate change.

4.7 New Zealand: Think Tanks and Astroturfing and Agriculture

This chapter has so far delved into the Australian CCCM. However, it is important to consider its neighbor, New Zealand. Despite being geographically close neighbors, the manifestation of CCCM in New Zealand was limited and appeared less influential on the surface. Nevertheless, CCCM organizations operated particularly between the late 1990s and 2000s, joining this global movement to undermine climate action and legislation.

One must look at the political and economic context to best understand the operations of New Zealand CCCM organizations and individuals. Much like the changes in the economy in the UK, in the 1980s, the New Zealand government began to transition its economy, selling off its largest government petroleum exploration company, increasing deregulation in the gas, petroleum, and electricity market, and facilitating an electricity market-based system. These issues were previously held under the Ministry of Energy before key elements were transferred to the Ministry of Commerce (New Zealand Government, 1997). Therefore, during the 80s to the early 1990s, it was clear that the influence of a neoliberal market had started to influence the energy and electricity markets of New Zealand, reflecting an environment through which the CCCM could easily find opportunities and targets. The evolution of "Rogernomics" under a Labor government followed and aligned with Reaganomics and Thatcherism in developing and deploying a market-led New Zealand monetary policy (Walker, 1989) and accompanying the success of the economic shift and policy changes would ultimately be think tanks. Goldfinch and Roberts (2013) reported on networks of neoliberal think tanks in Victoria (Australia) and New Zealand, which would provide a source of influence over government policy and directions with interlocking individuals revolving between businesses, governments, and think tanks. As such, the conditions of a think tank model that appeared in the CCCM, which was operational in Europe and the USA, would also manifest in New Zealand. Notably, the role of CCCM organizations was to frame the issue of climate change as primarily economic and combined with environmental.

One example was the public policy think tank, the New Zealand Initiative. Formed in 2012 as a merger of the New Zealand Business Roundtable and the New Zealand Institute, the think tank covers various policy issues, including poverty and protection from environmental problems. It has previously co-sponsored the Heartland Institute's International Conference on Climate Change. Of note here is that before its merger, the New Zealand Business Roundtable can best be described as an early variation of a CCCM organization. The organization was founded in 1986 with Roger Kerr as the executive director, who followed the ideological trends established under Rogernomics. Kerr was also a senior figure in the New Zealand Treasury and the Ministry of Foreign Affairs before his tenure at the Roundtable (Migone, 2011).

In a cartoon piece by Thomas Scott (1992), Kerr, the then director of Roundtable, is portrayed in a typical image associated with climate change denial and his "head in the sand." This image imitated the assumption that Kerr would avoid advocating for appropriate legislation on climate change at all costs. In fact, in 2007, the Roundtable hosted a lecture with contrarian Nigel Lawson followed by a discussion with Robert Kerr (Lawson, 2007). Kerr combined his ideological positions with discourses of delay (e.g., Kerr, 2010). In a group of essays published between 2002 and 2007 by the Roundtable, Kerr turned to the issue of climate change (Kerr, 2008). While he claimed to leave "aside the scientific uncertainties about global warming..." (p. 19), the discussion includes a key point of note:

> *The plan also involves further takings of property rights without compensation (for example, by penalizing farmers who change land use). It fails to recognize the property rights of pre-1990 foresters and the past efforts of businesses to reduce emissions.* (p. 20)

Kerr challenged the actions of the labor government and a series of policy decisions using cases from India, China, and the USA that had questioned and resisted implementing the Kyoto Protocol. While not denying the science of climate change, it reflects the well-known delaying tactics and the creation of doubt.

Kerr would go on to write a position for the New Zealand Center for Political Research (NZCPR). Founded in 2005, the NZCPR was led by Dr. Muriel Newman (NZCPR, n.d.-a), a former New Zealand politician, president of the Chamber of Commerce, and member of the ACT party. The ACT party was best considered a right-of-center party based on the Association of Consumers and Taxpayers. Thus, central to the party was advocating policies that expand personal freedoms and responsibility and increase deregulation. Sir Roger Douglas founded the NZCPR (NZCPR, n.d.-b). Douglas formed part of the parliament during the era of Rogernomics, putting his ideas into the Association of Consumer and Taxpayers as a pressure group before creating the ACT political party. The political party aligned with some elements of the Republican approach to climate change in the USA, emphasizing a critique of collective action in response to the climate crisis with reduced to no government intervention. Instead, intervention should be left to businesses, communities, and individuals (ACT, n.d.). It is unsurprising then that this position would later manifest in the NZCPR. The organization, since its orientation, consistently challenged conventional positions, including critiquing the science of climate change. At the same time, they would promote the work of Nigel Lawson (2014), Bob Carter (2008), David Bellamy (2007), James Delingpole (2010), and Ross McKitirick (2009). This included republishing the work of others and events such as Lawson's lecture from the New Zealand Business Roundtable.

Generally, these individuals who had also been associated with organizations outside of Australia helped diffuse information through CCCM organizations across countries. As described in previous chapters, we can trace many of these to the USA and Europe. Alongside individuals, there were also transnational networks such as the Climate Realist Network (Chapter 3). These transnational networks also reached New Zealand. This is best articulated in the case of the fringe-like group The New Zealand Climate Science Coalition (NZCSC). The NZCSC was associated with the International Climate Science Coalition (ICSC) (see Chapter 5), the Climate Science Coalition of America, and the Australia Climate Science Coalition (ACSC). Launched in 2006, the Coalition claimed to:

> *To represent accurately, and without prejudice, facts regarding climate change; to provide considered opinion on matters related to both natural and human-caused climate effects; and to comment on the economic and socio-political consequences of climate change.* (n.d., n.p.)

The organization was specifically created to focus solely on the issue of climate change in New Zealand. Still, it would reproduce some of the content from other associated organizations in other countries. In essence, the NZCSC would adapt to the political-economic conditions that shaped New Zealand following the practice of related countermovement organizations in its network. This overlap with existing CCCM organizations was cemented by its founders Professor August H. (Augie) Auer jr, David Bellamy, Bob Carter, and others (NZCSZ, n.d.). Honorary Secretary Terry Dunleavy and several other coalition members went on to sign the Dutch Clintel World Climate Declaration (Clintel, n.d.).

Over time, the NZCSC would target the New Zealand government's and international agencies' operations and actions. For example, in 2006, it claimed that the Ministry for the Environment was misleading New Zealand children about human-caused climate change and its impacts, using the argument presented by Auckland University professor and contrarian Chris De Frietas. In addition, in 2006, Bryan Leyland, Chairman of the NZCSC, released a press release that critiqued the Kyoto Protocol claiming: "Even if man-made global warming is happening, the IPCC climate models predict that adherence to Kyoto will not make a noticeable difference" (Leyland, 2006, n.p.). Moreover, the organization republished the segment from contrarian Marc Morano on several occasions to promote the same messages (e.g., NZCSC, 2007) that would continually critique the media alongside stories that would align with countermovement opposition.

Another fringe-like group was the Climate Realists (New Zealand). It was an advocacy organization dedicated to the topic of climate change. Founded in 2009, It provided a fact sheet (Climate Realists, 2009a) and sold numerous climate contrarian texts on its website (Climate Realists, 2009b). Central to this group was its founding by those in the "beef and sheep farming" industry (Climate Realists, 2009c, n.p.). Notably, the agricultural sector was one of the most significant contributors to New Zealand's overall emissions (New Zealand Government, 2021), with targets from the Kyoto Protocol and international governance strategies targeting reductions in emissions that would ultimately hit those industries. Therefore, unlike the fossil fuel industry's key role in Australian countermovement opposition, the agricultural sector became a central focal point for New Zealand CCCM organizations.

In response to limited bipartisan support for the development of the New Zealand Emissions trading scheme (NZ ETS) to meet New Zealand's climate goals (Bullock, 2012), the organization would claim:

> *We believe that mankind's activities have a minimal effect on the world's climate...the cost of an ETS is unsustainable...the introduction of an ETS in New Zealand may attempt to make us look good to our export markets, but it will have no effect whatsoever on the climate, and it will destroy New Zealand's agricultural base. For an example of a country which commits agricultural suicide, look no further than Zimbabwe.* (Climate Realists, 2009d, n.p.)

They would further promote the work of individuals such as David Bellamy (2009) and republish work and critiques from mainstream news outlets such as the Wall Street Journal (Climate Realists, 2009e). Moving forward, they would continue to expand their discussions critiquing scientists, including republishing the work of the Australian Galileo Movement (Galileo Movement, 2012d) and the Carbon Sense Coalition (e.g., Carbon Sense Coalition, 2010b). Moreover, much like the Australian Galileo Movement, they sent letters to politicians (Climate Realists, 2011) in attempts to engage in discussions and influence approaches to policy. However, evidence of their effectiveness was limited, to say the least.

The NZ ETS became law in 2008, set out under the 2002 Climate Change Response Act to meet the country's obligations under the Paris Agreement. Notably, the agricultural sector had some exclusions under the scheme based on particular regulatory positions outlined in the agreement. As Bullock (2012) noted, "heavy lobbying by the agriculture sector has in many ways been a distraction, enabling lobby groups to point to New Zealand's distinctiveness as a means of dismissing attempts to implement effective policy" (p. 659). This is unsurprising given the complex makeup of New Zealand's emissions, where land use changes and agricultural industry-related emissions were far more than other developed nations. Thus, rather than fossil fuels, lobbying from the agricultural sector has been a primary driver of opposition to implementing significant climate-related policy in New Zealand. Nevertheless, the agricultural-related CCCM organizations mirrored the efforts of those in other countries. Moreover, the individual overlaps with Australian CCCM organizations illustrate the diffusion of ideas and practices in what speculatively appears as an organized global campaign.

An example of such lobbying efforts and what in some way could be considered the CCCM was the industry association the Federated Farmers (FFNZ). FFNZ was an advocacy group representing farmers and their industry nationwide (FFNZ, n.d.). Like the New Zealand

Chamber of Commerce, FFNZ was an influential lobbying organization representing business and successfully accessing the political process (Strong & Tyler, 2017). Liepins and Bradshow (1999) described the history of the FFNZ in policymaking, outlining that the organization echoed "*important discursive motifs of neo-liberal economic reform as it attempts to recreate itself as a party with influence in national politics*" (p. 564). The New Zealand government's positions and FFNZ outputs emphasized the success of agribusiness deregulation. By emphasizing the success of a market-driven system as a reason to continue the trend of shaping policy under the neoliberal ideal which swept across the world from the 1980s, it would position and facilitate changes in the agribusiness industry that had previously been managed and supported by the government with accompanying regulations.

Using the success of this promotion of economic ideas and the transition towards a more market-driven system, the FFNZ would consider climate change and increased regulation as the issue. This was evidenced in 2007 in the organization's response to the Kyoto Protocol, claiming:

> *Federated Farmers opposed the government's 2002 ratification of the Kyoto Protocol and remains opposed to it...Our members are of the strong view that it is unjustifiable and inappropriate for the government to bind New Zealand to emissions reduction commitments and certain economic hardship for little if any environmental benefit.*

Over time climate change remained a vital issue for the organization. In their 2011 manifesto, they highlighted the concerns of its members and particularly the New Zealand ETS system, outlining that:

> *...Competitors in other countries have similar schemes or otherwise face similar emissions costs ...Climate change policies should be based on good* science, *be practical and cost-effective, and allow New Zealand farming to the construction of policy briefs.* (FFNZ, 2011. n.p.)

The FFNZ made clear the impacts that responding to climate change would have on its industry, reflecting similar responses to those of trade associations in the US fossil fuel sector. While not necessarily denying the science of climate change, the immediate reactions focused on delaying legislation for economic and financial viability and the uptake and development of mitigation technologies. Thus, agribusiness, much like the coal industry in Australia, played a similar role in the social, economic, and cultural makeup of New Zealand. As a result, the activities of these industries and representatives, such as the FFNZ, indicated a direct campaign to delay action by undermining and challenging government-based policies that sought to limit GHG emissions.

4.8 Conclusion

CCCM opposition in Australia can best be described as a network of think tanks, advocacy groups, industry associations, media organizations, PR campaigns, and individuals that operated to undermine climate action in Australia. While arguably not as significant and far-reaching in Australia as in the USA, the implications of the CCCM activities in Australia do appear to stand out, particularly their connections with the Tony Abbott administration before moving and finding their feet under the Scott Morrison administration. The movement may have impacted Australia's delayed action on climate change by relaying to the public and politicians that coal and gas were central to Australia's political economy. In New Zealand, central to the rise and operations of the CCCM is the crucial role of think tanks and their intersection with governments, much like in Europe. Alongside the focus on the agricultural sector, which contributed one of the largest amounts to the emissions in New Zealand. Ultimately, their industry would, in comparison to fossil fuels, be most at risk. Thus, it would mean some activities and focus shifted from fossil fuels to connections with the sector. While ultimately, the activities of the CCCM in New Zealand are less clear and have limited impact, it is not to say that they did not operate to undermine progressive legislation. Moreover, since 2012, the Climate Action Tracker's analysis of New Zealand indicated an insufficient response, especially in response to its first Kyoto Protocol targets. It is fair to say that there was clear evidence of countermovement opposition in Australia and New Zealand. Nevertheless, the foci of opposition organizations sometimes looked different, reflecting the pluriverse of countermovement opposition across other countries.

References

ABC. (2007). *Mark O'Neill: Australian coal*. https://www.abc.net.au/radionational/programs/archived/perspective/mark-oneill/3392944. Accessed 4 Oct 2022.

ACT Party. (n.d.). *Policies*. https://www.act.org.nz/policies. Accessed 4 Oct 2022.

Australian Climate Science Coalition. (2009). *About us*. https://web.archive.org/web/20120207065559/http://www.auscsc.org.au/about_us.html. Accessed 4 Oct 2022.

Australian Coal Association. (2003). *Publications, speeches and articles*. https://web.archive.org/web/20030622072003/http://www.australiancoal.com.au/publications.htm. Accessed 4 Oct 2022.

Australian Environment Foundation. (n.d.). *The AEF in the Australian media*. http://web.archive.org/web/20161014092818/http://aefweb.info/display/In_The_News.html. Accessed 4 Oct 2022.

Australian Environment Foundation. (2005). *Staff*. https://web.archive.org/web/20050623220333/http://aefweb.info/staff.php. Accessed 4 Oct 2022.

Australian Environment Foundation. (2006) *Inaugural conference 2006, caring for the environment in a changing world*. https://web.archive.org/web/20061029012422/http://www.aefweb.info/display/conference2006.html. Accessed 4 Oct 2022.

Australian Environment Foundation. (2007). *Staff*. https://web.archive.org/web/20070715224038/http://www.aefweb.info/staff.php. Accessed 4 Oct 2022.

Australian Environment Foundation. (2008). *AEF 2008 annual conference program*. https://web.archive.org/web/20090611154040/http://aefweb.info/display/speakers2008.html. Accessed 4 Oct 2022.

Beder, S. (n.d.). *Institute of public affairs*. Business Managed Democracy. https://www.herinst.org/BusinessManagedDemocracy/government/national/IPA.html. Accessed 4 Oct 2022.

Bellamy, D. (2007). *The global warming myth*. New Zealand Center for Political Research. https://www.nzcpr.com/the-global-warming-myth/. Accessed 4 Oct 2022.

Bellamy, D. (2009). Professor David Bellamy doubts man-made climate change. https://web.archive.org/web/20090927112149/http://climaterealists.org.nz/node/16. Accessed 4 Oct 2022.

Bert Kelly Research Center. (2011). *Booklaunch, 14 December 2011*. http://www.bertkelly.org.au/events.html. Accessed 4 Oct 2022.

Bloomberg. (n.d.). *Viv Forbes: Profile*. https://www.bloomberg.com/profile/person/6070219?leadSource=uverify%20wall. Accessed 4 Oct 2022.

Bullock, D. (2012). Emissions trading in New Zealand: Development, challenges and design. *Environmental Politics, 21*(4), 657–675. https://doi.org/10.1080/09644016.2012.688359

Carbon Sense Coalition. (n.d.-a). *Further reading*. https://carbon-sense.com/further-reading/. Accessed 4 Oct 2022.

Carbon Sense Coalition. (n.d.-b). *Climate smart*. https://carbon-sense.com/climate-smart/. Accessed 4 Oct 2022.

Carbon Sense Coalition. (2007a). *Look before you leap: A submission to 'climate smart'* (a proposed energy policy for Queensland). http://carbon-sense.com/wp-content/uploads/2007a/06/look.pdf. Accessed 4 Oct 2022.

Carbon Sense Coalition. (2007b). *Introducing the carbon sense coalition*. http://carbon-sense.com/wp-content/uploads/2007b/06/carbon-coalition.pdf. Accessed 4 Oct 2022.

Carbon Sense Coalition. (2010a). *Coal combustion*. https://web.archive.org/web/20130918160318/http://carbon-sense.com/wp-content/uploads/2010/06/coal-combustion.pdf. Accessed 4 Oct 2022.

Carbon Sense Coalition. (2010b). "*Earth hour earth hour earth hour or blackout night? Or blackout night? Or blackout night?*" https://www.climaterealists.org.nz/node/316. Accessed 4 Oct 2022.

Carbon Sense Coalition. (2015). *Paris conference threatens our access to energy*. https://web.archive.org/web/20160122130424/http:/carbon-sense.com/2015/11/03/access-to-energy/#more-1514. Accessed 4 Oct 2022.

Carter, R. M. (2007). *The myth of dangerous human-caused climate change*. https://web.archive.org/web/20080718201908/http://www.aefweb.info/data/Carter.pdf. Accessed 4 Oct 2022.

Carter, R. M. (2008). *2006 Climate change and government conference: Hansenism in the cause of command and control climate politics*. Research Report New Zealand Center for Political Research. https://www.nzcpr.com/test-post-68/. Accessed 4 Oct 2022.

Center for Independent Studies. (n.d.). *History*. https://www.cis.org.au/about/history/. Accessed 4 Oct 2022.

Center for Independent Studies. (1996). *The 1996 John Bonython lecture*. https://web.archive.org/web/19970103234246/https://www.cis.org.au/. Accessed 4 Oct 2022.

Center for Independent Studies. (1997). *News*. https://web.archive.org/web/19980420132439/http://cis.org.au/. Accessed 4 Oct 2022.

ChemEurope. (n.d.). *Australian industry greenhouse network*. https://www.chemeurope.com/en/encyclopedia/Australian_Industry_Greenhouse_Network.html. Accessed 4 Oct 2022.

Christoff, P. (2005). Policy autism or double-edged dismissiveness? Australia's climate policy under the Howard government. *Global Change, Peace & Security, 17*(1), 29–44.

Christoff, P. (2022). *Mining a fractured landscape: The political economy of coal in Australia.*

Climate Action Tracker. (n.d.). *Australia.* https://climateactiontracker.org/countries/australia/. Accessed 4 Oct 2022.

Climate Action Tracker. (2018). *New Zealand.* https://climateactiontracker.org/media/documents/2018/4/CAT_2012-06-25_CountryAssessment_NewZealand.pdf. Accessed 4 Oct 2022.

Climate Realists. (2009a). *Climate change: The facts.* http://www.climaterealists.org.nz/node/603. Accessed 4 Oct 2022.

Climate Realists. (2009b). *Bumper stickers.* http://www.climaterealists.org.nz/node/352. Accessed 4 Oct 2022.

Climate Realists. (2009c). *About us.* https://www.climaterealists.org.nz/node/5. Accessed 4 Oct 2022.

Climate Realists. (2009d). *Homepage.* https://web.archive.org/web/2009d0922225013/https://www.climaterealists.org.nz/node/1. Accessed 4 Oct 2022.

Climate Realists. (2009e). *Wall Street Journal rebukes NZ on its climate variation response.* https://web.archive.org/web/2009e0924144127/http://climaterealists.org.nz/node/12. Accessed 4 Oct 2022.

Climate Realists. (2011). *Letters.* https://web.archive.org/web/20111125102737/http://www.climaterealists.org.nz/taxonomy/term/3. Accessed 4 Oct 2022.

Clintel. (n.d.). *World climate declaration: New Zealand.* https://clintel.org/new-zealand-wcd

Community Information Unit of the Department of the Environment Sport and Territories. (1994). *Climate change: Australia's national report under the United Nations framework convention on cimate change.* https://unfccc.int/sites/default/files/resource/ausnc1%20Australia.pdf. Accessed 4 Oct 2022.

Counterpoints. (2009). *No title.* https://web.archive.org/web/20130531085054/http://mpegmedia.abc.net.au/rn/podcast/2009/11/cpt_20091130_1605.mp3. Accessed 4 Oct 2022.

Crowley, K. (2021). Fighting the future: The politics of climate policy failure in Australia (2015–2020). *Wiley Interdisciplinary Reviews: Climate Change, 12*(5), e725.

Curran, G. (2021). Coal, climate and change: The narrative drivers of Australia's coal economy. *Energy Research & Social Science, 74*, 101955.

De Freitas, C. (2007, September 8–9). *Perspectives on global warming science.* [AEF Conference Melbourne]. https://web.archive.org/web/20080718201523/http://www.aefweb.info/data/DeFreitas.pdf. Accessed 4 Oct 2022.

Delingpole, J. (2010). *Global cooling and the new world order*. https://www.nzcpr.com/test-post-178/. Accessed 4 Oct 2022.

Department of Foreign and Trade Affairs. (2001, May 9). *Address on climate change to the Australian institute of international affairs by Mr. Ralph Hillman, Ambassador for the environment*. https://www.dfat.gov.au/news/speeches/Pages/address-on-climate-change-to-the-australian-institute-of-international-affairs-by-mr-ralph-hillman-ambassador-for-the-envir. Accessed 4 Oct 2022.

Evans, R. (2000). *Moving beyond Kyoto: A responsible approach to climate change*. http://www.lavoisier.com.au/articles/climate-policy/politics/evans2000-3.php. Accessed 4 Oct 2022.

Evans, R. (2002, January 24). The greenhouse-warming debate is hotting up. *The Canberra Times*. http://www.lavoisier.com.au/articles/climate-policy/politics/evans2002-1.php. Accessed 4 Oct 2022.

Evans, R. (2007). *The IPCC and AR4*. http://www.lavoisier.com.au/articles/greenhouse-science/climate-change/Evansonlineopinionfeb2007.pdf. Accessed 4 Oct 2022.

Federated Farmers. (n.d.). *About us*. https://www.fedfarm.org.nz/FFPublic/FFPublic/About_Federated_Farmers.aspx?hkey=e30f8cb6-f377-42a9-89fb-3dd43c009206. Accessed 4 Oct 2022.

Federated Farmers. (2011). *2011 Manifesto*. https://web.archive.org/web/20120804194317/http://www.fedfarm.org.nz/f1738,125264/125264_Federated_Farmers_Manifesto_2011.pdf. Accessed 22 Nov 2022.

Fielding, K. S., Head, B. W., Laffan, W., Western, M., & Hoegh-Guldberg, O. (2012). Australian politicians' beliefs about climate change: Political partisanship and political ideology. *Environmental Politics, 21*(5), 712–733.

Foyster, G. (2019, April 24). How Abbott still haunts climate policy. *Eureka Street*. https://www.eurekastreet.com.au/article/how-abbott-still-haunts-climate-policy. Accessed 4 Oct 2022.

Galileo Movement. (n.d.). *Who we are*. http://www.galileomovement.com.au/who_we_are.php. Accessed 4 Oct 2022.

Galileo Movement. (2011). Two weeks of thunder—Letters to several senators. https://web.archive.org/web/20150429024746/http://galileomovement.com.au/blog/. Accessed 4 Oct 2022.

Galileo Movement. (2012a). *Homepage*. https://archive.ph/9sN7M#selection-303.0-307.74. Accessed 4 Oct 2022.

Galileo Movement. (2012b). *The restoration of freedom, democracy, sovereignty & scientific truth in Australia—Identifying the failings and moving forward*. http://www.galileomovement.com.au/australia_democracy.php. Accessed 4 Oct 2022.

Galileo Movement. (2012c). *Tours*. https://web.archive.org/web/2012b0702222313/http://www.galileomovement.com.au/tours.php. Accessed 4 Oct 2022.

Galileo Movement (2012d). *The cost to you—In a nutshell*. http://www.galileomovement.com.au/cost_nutshell.php. Accessed 4 Oct 2022.

Global Energy Monitor Wiki. (n.d.). *Ralph Hillman*. https://www.gem.wiki/Ralph_Hillman. Accessed 4 Oct 2022.

Goldfinch, S., & Roberts, V. (2013). New public management and public sector reform in Victoria and New Zealand: Policy transfer, elite networks and legislative copying. *Australian Journal of Politics & History, 59*(1), 80–96.

Hamilton, C. (2009). *Meet the new dirty dozen*. Crikey. https://www.crikey.com.au/2009/06/16/greenhouse-mafia-revisted-meet-the-new-dirty-dozen/. Accessed 4 Oct 2022.

Hogbin, G. (1998). *Global warming: The mother of all environmental scares*. Center for Independent Studies.https://web.archive.org/web/19981212023806/https://www.cis.org.au/. Accessed 4 Oct 2022.

Hurcombe, E. (2016). The making of a captain: The production and projection of a political image on the Tony Abbott Facebook page. *Communication, Politics & Culture, 49*(1), 19–38.

Humphreys, J. (2007). *Exploring a carbon tax for Australia*. Center for Independent Studies Policy Monograph. Center for Independent Studies. https://www.cis.org.au/wp-content/uploads/2015/07/pm80.pdf. Accessed 4 Oct 2022.

Institute of Public Affairs. (n.d.). *About*. https://ipa.org.au/about. Accessed 4 Oct 2022.

Institute of Public Affairs. (1968). About the IPA. *IPA Review, 22*(2). https://ipa.org.au/wp-content/uploads/archive/1210645977_document_review1968_vol.22no.2.pdf. Accessed 4 Oct 2022.

Institute of Public Affairs. (2012). *Only 1/3 of Australians believe humans are to blame for global warming*. https://ipa.org.au/wp-content/uploads/archive/1332888123_document_280312_media_release_-_only_one_third_of_australians_believe_humans_are_to_blame_for_global_warming.pdf. Accessed 4 Oct 2022.

Kelly, C. R. B. (1978). *One more nail*. AbeBooks.

Kerr, R. (2008). *Do economists agree on anything and other essays?* New Zealand Business Roundtable. ISBN 978–1–877394–17–1.

Kerr, R. (2010). *Global warming no longer cool even in New Zealand*. New Zealand Centre for Political Research. https://www.nzcpr.com/test-post-152/. Accessed 4 Oct 2022.

Lavoisier Group. (2000). *Submission to the senate references committee for the environment, communications, information technology and the arts the renewable energy (electricity) bill renewable energy (electricity) (charge) bill 2000 introduction.* http://www.lavoisier.com.au/articles/greenhouse-science/climate-change/SenateSubmission1.pdf. Accessed 4 Oct 2022.

Lavoisier Group. (2002). *Submission to the JSCOT inquiry into the Kyoto protocol.* http://www.lavoisier.com.au/articles/climate-policy/science-and-policy/JSCOT.pdf. Accessed 4 Oct 2022.

Lawrence, P. (2009). Australian climate policy and the Asia Pacific partnership on clean development and climate (APP). From Howard to Rudd: Continuity or change?. *International Environmental Agreements: Politics, Law and Economics, 9*(3), 281–299.

Lawson, N. (2007). *A cool look at global warming: The economics and politics of climate change.* New Zealand Business Roundtable. Wellington.

Lawson, N. (2014). *Cool it.* New Zealand Center for Political Research. https://www.nzcpr.com/cool-it/. Accessed 4 Oct 2022.

Leviston, Z., Leitch, A., Greenhill, M., Leonard, R., & Walker, I. (2011). *Australians' views of climate change.* CSIRO.

Leyland, B. (2006, May 12). Kyoto and a carbon tax will make no noticeable difference in CO2 levels. *Scoop News.* https://web.archive.org/web/20060806114314/http://www.climatescience.org.nz/assets/2006512202690.KyotoC.TaxNoDifference.pdf. Accessed 4 Oct 2022.

Liepins, R., & Bradshaw, B. (1999). Neo-liberal agricultural discourse in New Zealand: Economy, culture and politics linked. *Sociologia Ruralis, 39*(4), 563–582.

Lomborg, B. (2006, August 6). The real priorities. *The Australian.* https://www.cis.org.au/commentary/opinion/the-real-priorities/. Accessed 4 Oct 2022.

Lucas, A. (2021). Investigating networks of corporate influence on government decision-making: The case of Australia's climate change and energy policies. *Energy Research & Social Science, 81*, 102271.

Lucas, A. (2022). *Fossil networks and dirty power: The politics of decarbonization in Australia.* Edward Elgar Publishing.

McGauchie, D. (2000). *Greenhouse and agriculture a paper delivered to the Lavoisier Group conference.* http://www.lavoisier.com.au/articles/climate-policy/science-and-policy/mcgauchie2000-c3.php. Accessed 4 Oct 2022.

MacNeil, R. (2021). Swimming against the current: Australian climate institutions and the politics of polarization. *Environmental Politics, 30*(supl), 162–183.

Maley, B. (2000). *Global warming raises political temperatures.* Commentary. https://www.cis.org.au/commentary/opinion/global-warming-raises-political-temperature/. Accessed 4 Oct 2022.

Mayne, S (2015). *Mayne: Which corporations donated to put Tony Abbott in the lodge?* Crikey. https://www.crikey.com.au/2015/02/02/mayne-which-corporations-donated-to-put-tony-abbott-in-the-lodge/. Accessed 15 Dec 2022.

McKitirick, R. (2009). *Defects in key climate data are uncovered*. https://www.nzcpr.com/defects-in-key-climate-data-are-uncovered/. Accessed 4 Oct 2022.
Mendes, P. (2003). Australian neoliberal think tanks and the backlash against the welfare state. *Journal of Australian Political Economy, 51*, 29–56.
Migone, P. (2011). *Tributes flow for business leader Roger Kerr*. Business Day. Stuff.co.nz. https://web.archive.org/web/20111029131629/http://www.stuff.co.nz/business/5875388/Business-leader-Roger-Kerr-dies. Accessed 4 Oct 2022.
Milman, O. (2011, May 31). How mining and media distort Australia's carbon tax debate. *The Guardian*. https://www.theguardian.com/environment/blog/2011/may/31/mining-media-australia-carbon-tax. Accessed 4 Oct 2022.
Ministry of the Environment. (n.d.). *About the New Zealand emissions trading scheme*. https://environment.govt.nz/what-government-is-doing/areas-of-work/climate-change/ets/about-nz-ets/. Accessed 4 Oct 2022.
Moran, A (eds). (2010). *Climate change: The facts*. https://ipa.org.au/wp-content/uploads/archive/1321487125_document_moran_climatechange-thefacts.pdf. Accessed 4 Oct 2022.
New Zealand Climate Science Coalition. (n.d.). *About us*. https://www.nzclimatescience.org/aboutus. Accessed 4 Oct 2022.
New Zealand Climate Science Coalition. (n.d.). *Homepage*. https://web.archive.org/web/20060719001305/http://www.climatescience.org.nz/. Accessed 4 Oct 2022.
New Zealand Climate Science Coalition. (2006). *Press release*. https://web.archive.org/web/20060806113506/http://www.climatescience.org.nz/assets/2006631540560.GreenhouseRelease.pdf. Accessed 4 Oct 2022.
New Zealand Climate Science Coalition. (2007). *310707 Marc Morano's round-up*. https://www.nzclimatescience.org/blog/310707%20marc%20moranos%20round-up. Accessed 4 Oct 2022.
New Zealand Government. (1997). *Energy use*. https://environment.govt.nz/publications/the-state-of-new-zealands-environment-1997/chapter-three-production-and-consumption-patterns/energy-use/. Accessed 4 Oct 2022.
New Zealand Government. (2021). *New Zealand greenhouse gas inventory 1990–2020*. https://environment.govt.nz/publications/new-zealands-greenhouse-gas-inventory-1990-2020-snapshot/. Accessed 4 Oct 2022.
New Zealand Center for Political Research. (n.d.-a). *About us*. https://www.nzcpr.com/about-us/. Accessed 4 Oct 2022.
New Zealand Center for Political Research. (n.d.-b). *Principles*. https://www.act.org.nz/principles. Accessed 4 Oct 2022.
Nicol, J. (2008, September 24). Experts in hot air? *The Australian*. https://web.archive.org/web/20081211170203/http://www.theaustralian.news.com.au/story/0,25197,24391980-21682,00.html. Accessed 4 Oct 2022.

Nova, J. (2009). *The skeptics handbook: Evidence of warming.*
https://joannenova.com.au. (n.d.). *Homepage.*
Parliament of Australia. (n.d.). *Hon Tony Abbott*. https://www.aph.gov.au/Senators_and_Members/Parliamentarian?MPID=EZ5. Accessed 4 Oct 2022.
Pearse, G. (2007). *High and dry, John Howard, climate change and the selling of Australia's future*. Viking, Penguin.
Pearse, G. (2011a). *The climate movement: Australia's patrons of climate change activism*. https://www.wrongkindofgreen.org/2011a/09/26/the-climate-movement-australias-patrons-of-climate-change-activism/. Accessed 4 Oct 2022.
Pearse, G. (2011b). *The climate movement: Australia's patrons of climate change activism*. The Wrong Kind of Green. https://www.wrongkindofgreen.org/2011b/09/26/the-climate-movement-australias-patrons-of-climate-change-activism/. Accessed 8 Oct 2022.
Plimer, I. (2011). *How to get expelled from school: A guide to climate change for pupils*. Connor Court Publishing Pty Ltd.
Readfearn, G. (2018, July 20). Gina Rinehart company revealed as $4.5m donor to climate skeptic thinktank. *The Guardian*. https://www.theguardian.com/business/2018/jul/21/gina-rinehart-company-revealed-as-45m-donor-to-climate-sceptic-thinktank. Accessed 8 Oct 2022.
Ridd, P. (2017). *The extraordinary resilience of the great barrier reef corals, and the problems with policy science*. http://web.archive.org/web/20180320191359/http://blackjay.net/wp-content/uploads/2018/01/Ridd-P-Chapter-1-from-Climate-Change-The-Facts-2017-IPA.pdf. In J. Marohasy (Ed.), *Climate change: The facts 2017* (pp. 9–23). Connor Court Publishing.
Rocklands Richfield Limited. (2006). *Bowen Basin expert joins Rocklands Richfield coal company Viv Forbes appointed general manager*. Media release. https://www.desmog.com/wp-content/uploads/files/Viv%20Forbes%20appt%2016%20Mar%2006.pdf. Accessed 8 Oct 2022.
Scott, T. (1992). Scott, Thomas 1947: The Business Roundtable deny the existence of global warming... ; Mind you, this sand is hot... ; Furthermore, the absolutely penniless are paid too much... Roger Kerr 30 June 1992. https://tiaki.natlib.govt.nz/#details=ecatalogue.106728
Source Watch. (n.d.). *Lavoisier Group*. https://www.sourcewatch.org/index.php/Lavoisier_Group. Accessed 4 Oct 2022.
Stanmore Coal. (n.d.). *Directors and management*. http://web.archive.org/web/20110407193124/www.stanmorecoal.com.au/corporate_directors_and_management.aspx. Accessed 4 Oct 2022.
Strong, C., & Tyler, F. (2017). New Zealand media camouflage political lobbying. *Pacific Journalism Review, 23*(2), 144–158.
Tobacco Tactics. (2020). *The institute of public affairs*. https://tobaccotactics.org/wiki/institute-of-public-affairs/. Accessed 8 Oct 2022.

Walker, J. (2021). Freedom to burn: Mining propaganda, fossil capital and the Australian neoliberals. In Slobodian and Plehwe (Eds.), *Market prophets from the margins: Neoliberals east and south*. Near Futures.
Walker, S. (1989). *Rogernomics: Reshaping New Zealand's economy*. GP Books.
White, A. (2014). *How Rupert Murdoch created the world's newest climate change villain Salon*. https://www.salon.com/2014/06/22/how_rupert_murdoch_created_the_worlds_newest_climate_change_villain/. Accessed 4 Oct 2022.
Wright, C., Irwin, R., Nyberg, D., & Bowden, V. (2022). 'We're in the coal business': Maintaining fossil fuel hegemony in the face of climate change. *Journal of Industrial Relations*. https://doi.org/10.1177/00221856211070632
Wright, C., Nyberg, D., & Bowden, V. (2021). Beyond the discourse of denial: The reproduction of fossil fuel hegemony in Australia. *Energy Research & Social Science, 77*, 102094.
Young, S. (2017). News corporation tabloids and press photography during the 2013 Australian federal election. *Journalism Studies, 18*(7), 866–889.

CHAPTER 5

Canada and Petro-Nationalism

5.1 Introduction

This chapter first describes the complex entanglements between the fossil fuel industry and the socially constructed, social, political, and cultural significance of fossil fuels in Canada. In so doing, it illuminates the conditions for which CCCM organizations in Canada would manifest, beginning with the role of the Fraser Institute. The role of the Fraser Institute in the CCCM was only minor compared to its overall success in shaping and facilitating the spread of neoliberal ideas in Canada. Interlocking with the institute, the chapter then explores the expanding characteristics of a think tank network before the rise of Stephen Harper's government. During his tenure, it became clear that the CCCM elevated its work by crafting more mainstream PR campaigns to reinforce the legitimacy of fossil fuels as central to the growth of the Canadian economy and the welfare and prosperity of the population.

5.2 The Hegemony of Fossil Fuels

Imperial Oil, the Canadian arm of Exxon Mobil, could have taken serious action to address the climate crisis. In the 1970s, the company launched a research program focused on the issue of climate change, demonstrating at least on the surface, some engagement on the subject of climate change outlining suggestions and proposals to address the problem. However,

R. E. McKie, *The Climate Change Counter Movement*,
https://doi.org/10.1007/978-3-031-33592-1_5

these apparent forward steps to address the industry's impact on the climate were entangled in a broader campaign that would undermine, delay, and obstruct climate action. More specifically, as research continued from the late 1990s, Imperial Oil's more positive potential contributions to reducing GHG emissions began to transform, incorporating similar discourses of delay to those observed in the US (ClimateInvestigations, n.d.).

To understand the operations of the CCCM in Canada, it is essential to explore the status of Canada as a first world Petro-state' (Nikiforuk, 2010). Fast (2014) emphasizes that extractive industries were central to Canada's political economy. And since its rise as an independent nation, central to its successful development was the extraction of natural resources, which continued and expanded following a shift under the global neoliberal regime. As a reminder, neoliberalism is more than simply an ideology. It shapes the values and belief systems of individuals, organizations, and institutions whereby capitalism and neoliberal variations became somewhat naturalized as if there were no alternatives. As Eaton and Day (2020) described, the hegemony of the oil and gas industry even begins in education. This is unsurprising given that some education projects are from corporate-funded non-profits. Thus, even from childhood, the centralization of oil and gas could be embedded in the worldviews of the Canadian population.

In simplistic terms, the ideology of neoliberalism problematizes state intervention on some level. As such, any policy developments which sought to reshape land use and property rights with First Nation lands would be challenged (Carroll, 2020; Piggot, 2018). Such an assessment is not unique to Canada. Historically and globally, despite shifts in domestic and international policy that sought to recognize the rights of land in areas occupied by indigenous peoples that required increased consultation and participation in decision-making, conflicts remained. This is despite perceptions and trends that technological advances, increased liberalization, and privatized corporate investment were increasing economic development in these locations (Whiteman & Mamen, 2002). Therefore, the combination of international agreements and targets that would impact Canadian domestic policy and corporate interests in expanding territories would delay significant action on climate change, despite its negative ecological and social consequences (e.g., Parson & Ray, 2016).

In Chapter 1, I clarified the role of fossil fuels in stimulating the global neoliberal capitalist economy. In so doing, I showed how this led to a series of complex social relationships where wealth became concentrated in the hands of a minority as it relates to the destruction of the environment. In Canada, it specifically manifests in the concentration of corporate ownership of fossil fuel companies. The corporate ownership of the Canadian fossil fuel industry was largely concentrated in the hands of 25 domestic and foreign companies. Between 2010 and 2015 these included Exxon Mobil Corp, Royal Bank of Canada, and Black Rock Inc (Carroll & Huijzer, 2018). Like the make-up of fossil fuel economies including the USA, concentration of wealth for corporate actors was essential, and influencing policy would become of primary interest.

In support of the fossil industry, think tanks, advocacy groups, and industry associations were central vehicles for disseminating discourses of delay reaching the media, transcending into the public voice and the political sphere in Canada. As Carroll et al. (2018, p. 426) described; "Corporate influence is, at its core, geared toward protecting investments and profit streams, opening new fields for investment, and minimizing intrusions into profit, such as taxes, regulations, and unions." These same ideological foundations, and political-economic forces driving the CCCM in Australia, USA, and Europe, also manifested within Canada. Carroll (2020) outlined distinct areas of corporate power that helped obstruct effective climate policies in Canada. This included business leadership, business-friendly think tanks, advocacy groups, foundations and political parties, PR firms, media echo chambers, and corporations. While the fossil fuel industry could directly donate to political parties, other groups were able to lobby and influence the political process and public opinion on the issue of climate change. Importantly, this chapter first turns to the role of think tanks central in maintaining fossil capitalism influence in Canada acting as one of arm of the CCCM.

5.3 The Fraser Institute (1989–2006)

Between 1993 and 2006, the Liberal Party dominated the Canadian political sphere. While the previous conservative government was described as a leader in action to mitigate rising GHG emissions, it was Jean Chretian, then Liberal Prime Minister, who ratified the Kyoto Protocol (Smith, 2009). After that, however, it became evident Canada would not only be unable to meet its Kyoto targets but was on a course to fail at the 1992 Rio conference. During this period, as the issue of climate change became a key focus of policy, think tanks in Canada mobilized the government's response to the problem. The most infamous of these is the Fraser Institute which, as was the case with US think tanks, received funding from donors including ExxonMobil ($120,000 between 2003 and 2004), Charles G. Koch Charitable Foundation ($1,669,721 between 1992 and 2018) and other conservative foundations (see Conservative Transparency, n.d.).

Headquartered in Vancouver, the institute was founded in 1974 by Michael A. Walker, an MPS member (Chapter 6), who remained Executive Director until 2004 before becoming a senior fellow. Critical to the founding of the Fraser Institute was Anthony Fisher (Chapter 3), who then became co-director of the institute. Like similar CCCM organizations, central to the positions of the Fraser Institute were neoliberal principles, expansion of markets, decreasing and limited regulation including environmental regulations, and a reduction in government intervention. A Canadian economist, Walker would go on to foster these ideas into the institute, disseminating policy positions aligned with this ideological thinking. Similarly, Fisher's connections illustrated the fundamental global diffusion of ideas that would in some element of the issue draw and then turn their particular attention to the issue of climate change. From the IEA in the UK, Center for independent Studies in Australia, and the Fraser Institute in Canada, the rise of think tanks in the political sphere would help legitimize climate oppositional positions.

The first clear sign of engagement on the climate issue was in 1989 when the Fraser institute released: *"Economics and the Environment: A Reconciliation."* This was an edited collection by Walter Block, with contributing authors including Richard Stroup (a previous member of the Heartland Institute's experts (Heartland, n.d.) and Murray Rothbard (Chapter 3) highlighting one of the first clear connections of the Fraser Institute and its delaying discourse embedded in the ideological undercurrents underpinning the CCCM observed in the USA, Europe, and Australia. The central argument proposed in the text was to promote free market environmentalism, with an emphasis on property rights and reduced government intervention (Block, 1990). This collection challenged significant developments within proposed national and global legislation. It starkly contrasted some arguments advocated by environmental movements asking for increased regulatory intervention to address climate change. More specifically, central to the discourse was the direct comparison to an individualistic and market-focused response as opposed to the collective action and increased state regulation proposed by environmental and climate justice movements. While both arguments can be legitimized as a strategy to address climate change, this alternative market-based approach offers a consistent pattern to likewise countermovement opposition in the USA, Europe, and Australia.

Over time, the institute remained vocal in its opposition. It became a member of the CHC (Chapter 2) and in late 1997, hosted an international conference that critiqued the a*pocalyptic* impacts of climate change with the proceedings later released; *Global Warming: The Science and the Politics* (Jones, 1997). During the early 2000s, the institute had a dedicated section on the environment, which often covered the issue of climate change (e.g., Fraser Institute, 2002a, 2002b, 2022c). In addition, the section would present editorials and positions from members of the organizations that included discourses of delay. For example, in 2002, the institute held the event *Taken By Storm: The Troubled Science, Policy and Politics of Global Warming*. In 2004, they promoted the work of Andrei Illarionov, Russian libertarian economist, former economic policy advisor to Vladimir Putin, and former senior fellow at the Cato Institute (McKie, 2021), who had framed climate change policy concerning the "national interest" (Tynkkynen, 2010). In this case, Illarionov critiqued climate science and the Kyoto Protocol by comparing Russia's resistance to the treaty as an example of what Canada should do (Illarinov, 2004). This event followed a meeting organized by the Russian Academy of Sciences (RAS), where contrary to what could have been a productive meeting, instead gave weight to the voices of skeptics. This further legitimized Iiarinov's position that the Kyoto Protocol would undermine economic growth in Russia and stall progress (Enserink, 2004). This critique and justification for not signing the Kyoto Protocol was reflected in the same positions on Canada's entry into the Protocol articulated in his speech at the Fraser Institutes' event. However, we cannot and should not say that Illarinov's positions influenced these actions.

Similarly, following the 2007 UNFCCC report, the institute released *The Independent Summary for Policymakers* claiming to provide a detailed and balanced overview of the 2007 IPCC report (2007). However, throughout the text, there were chapters based on disputing the scientific evidence of human-caused climate change with the final section of the summary outlining:

> *There is no evidence provided by the IPCC in its Fourth Assessment Report that the uncertainty can be formally resolved from first principles, statistical hypothesis testing or modeling exercises. Consequently, there will remain an unavoidable element of uncertainty as to the extent that humans are contributing to future climate change, and indeed whether such change is a good or bad thing.* (2007, p. 8)

Attending the event for the physical release of the report was Illarinov and Ross McKitrick and the event was likewise premised on challenging the findings from the UNFCCC. In 2014 Kenneth Green, then senior fellow created the report: *Climate Change: It's Not the Science, It's the Policy*. The focus was a critique of the UNFCCC and climate governance, questioning the economic impacts of policies. Green was a former scholar at the American Enterprise Institute (AEI) (n.d.) and had previously signed a joint letter with the CEI (including Myron Ebell) to the Bush administration intending to support the removal of the USA from the Kyoto Protocol (Ebell & Smith, 2002). Other supporters of the letter included David Rothbard (CFACT, USA), Fred Singer (SEPP, USA) and Paul Driessen (Center for the Defense of Free Enterprise, USA).

The letter notes partisanship: "We urge you to dismiss or re-assign all administration employees who are not pursuing your agenda, just as you have done in several similar instances" (n.p.). Targeting US politicians, the letter's authors emphasized the importance of partisanship facilitating climate opposition and its association with the Republican Party. However, the same scenario did not necessarily apply in the case of Canada and was more complex. This is because the Canadian political system was historically less partisan, whereby the two main parties often had some similarities and the same interests. This limited the opportunities for more ideological extremes with usually very low political activism from the public (Cochrane, 2010).

On the one hand, the issue of climate change and policy responses were somehow divided across party lines. On the other hand, the issue of climate change was somewhat depoliticized. Instead, because fossil fuel extraction was tied to Petro-nationalism, it would impact both sides of the political spectrum (Gunster et al., 2021). Even so, as Gunster et al. (2021) note, there were active attempts on the political left to mobilize against continued extraction, which ultimately reflects the conditions for which CCCM opposition can foster. In other words, "countermovements cannot develop without the presence of a viable threat, as countermovements exist to protect the interests of the dominant class and the status quo" (Wrenn, 2022, p. 243). As such, these organizations with a broader focus on policy areas following a particular ideological alignment would use their legitimized positions in the policy realm as think tanks to advocate for existing practices that did not reflect the long-term impacts of fossil fuel exploitation under a neoliberal capitalist economy.

To help further the critiques on climate justice movements and progressive legislation the Fraser Forum acted as a vehicle to emphasize the positives of the fossil fuel industry. The Fraser Forum is the institute's blog that accrued many posts under an "environment" subsection (Fraser Institute, n.d.-a, n.d.-b.). The blog promoted posts that included a host of discourses of delay, even emphasizing the positives of fossil fuels for the environment and the economy. For instance, the organization reposted Kenneth Green's appearance on a mainstream news network CBC (Green, 2014a, 2014b), claiming that oil pipelines are "*safer, cheaper, and greener*" (Green & Jackson, 2015, n.p.) and promoting fossil fuels and independent car use over public services such as rail transport (Green, 2015). Furthermore, promoting independent car use reinforced the frame that climate change could be solved by shaping individual behavior over introducing public services and government intervention. In essence, it was an argument that helped reinforce the belief system to sustain and protect neoliberal capitalism and importantly, the role of fossil fuels.

Like other forms of media, the blogosphere elevated the voices of CCCM opposition (Sharman, 2014). Sharman (2014) reports on the role of nine blogs based in Canada and the global climate skeptic Blogosphere. They explain that these blogs are largely "preoccupied with framing climate change as an active scientific issue" (p. 167). This approach to promoting science denial on the outskirts of the movement was useful because it helped remove elements of presumed political ideology often attached to the issue of climate change in political discourse and dominant in policy discussions and reports by think tanks such as the Fraser Institute. One case is the blog *Climate Audit* whose founder and editor, Stephan McIntyre, had a history of working for mining corporations, including President at Northwest Explorations. He later became part of CGX Energy when he became a strategic advisor (CGX Energy Inc, 2003). Thus, McIntyre operated within the CCCM and held direct ties to the fossil industry in Canada. Furthermore, the Fraser Institute promoted the work of and activities of Stephan McIntyre like others who adopt oppositional positions to the mainstream scientific consensus (e.g., Schneider, 2006), demonstrating an example of interlocking relationships between fossil fuel interests and countermovement opposition. In turn, think tanks could maintain a policy focused *unbiased* [emphasis added] position, while the scientific skepticism, over time became problematic and significantly undermined because of the growing evidence, was side-lined but still operationalized by interconnected members of the blogosphere (see also Coan et al., 2021 for more details on the blogosphere and think tank connection).

5.4 Think Tanks, Ideology and Expanding Opposition

Several other think tanks operated—intentionally or unintentionally—as opposition movement in Canada, with some funded by large corporations that operated throughout the Liberal party's leadership and the transition into Stephen Harper government. These included the C.D. Howe Institute which maintained a high profile in discussions on policy (Carroll et al., 2018). According to Carroll et al. (2018) the institute was one of the biggest funded corporate organizations in the *Carbon Capital Network*. In what appears to be one of the earliest available outputs, the institute promoted a piece authored by Daniel Schwanen questioned climate science and criticized policies that would harm the economy:

> *In devising ways to reduce emissions of greenhouse gases (GHGs), Canada's policy toward climate change should neither cause unnecessary declines in Canadians' standard of living nor put Canada at an unfair disadvantage relative to other countries... there is as yet no scientific consensus on just how serious a problem heat-trapping GHGs caused by human activity might be....* (1997)

The institute continued to disseminate material including a response to the Kyoto Protocol (e.g., Doucet, 2004) with a position written by Ross McKitrick and Randall Wigle (2002). The authors emphasized the economic costs of ratifying the Kyoto Protocol. In addition, Carroll et al. (2018) identified ties between the C.D. Howe Institute and the Canadian Energy Pipeline Association (CEPA) industry group. The CEPA represented 97% of Canada's oil and natural gas pipeline companies

(Corporate Mapping, n.d.-a). While CEPA would promote their support for addressing climate change, and included the note in an annual report:

> *It was also noted that greenhouse gas emissions should be reduced in all-consuming and producing sectors of the economy through energy efficiency practices and not through reductions in Canadian industry output for domestic or export markets.* (CEPA, 2003, n.p.)

Of interest is the positive association from CEPA to addressing the climate crises yet activating the messages of a Petro-nationalism deeply rooted in the Canadian political economy.

The Frontier Center for Public Policy (FCPP) was a think tank focused on various policy issues (FCPP, n.d.). It operated a specific research area on the environment, energy, and climate that, over time, cited the research of familiar individuals that had espoused discourses of delay including Paul Driessen, Patrick Moore, and Willie Soon. In 1997 they promoted a commentary by Peter Holle which focused on the questionable economic impacts and harms of addressing "global warming" and critiquing bad policies because of unproven beliefs (Holle, 1997, n.p.). Later, in 2002, they promoted a newspaper report by David E. Wojick and the text *A Skeptics Guide to Global Warming* (Frontier Center, 2002). David E. Wojick was a collaborator with the Heartland Institute (Heartland Institute, n.d.) and an adviser to the GES (Chapter 2). Moreover, the FCPP was a member of the CSCCC (Chapter 6). While it remains unknown if and what fossil fuel related funding the Frontier Center may have received, these indirect ties and the promotion of discourses of delay by others indicated its role in the broader CCCM in Canada.

A final example is the network of think tanks under the Energy Probe Research Foundation (EPRF). The organization openly received donations from the Atlas Economic Research Foundation (Chapter 3). It was separated into five divisions; *Energy Probe, Probe International, Environment Probe, Consumer Policy Institute, Urban Renaissance Institute and Environmental Bureau of Investigation.* Very much operating as a front group with claims of 'conservation' and environmental protection, the Energy Probe division claimed:

> *The consumer and energy research team at EPRF active in the fight against nuclear power, and dedicated to resource conservation, economic efficiency,*

and effective utility regulation. Most recently, with Lawrence Solomon's blockbuster book, The Deniers, Energy Probe has led the charge that the science of man-made global warming is not settled and defended those scientists who work to enlighten and inform the debate.

Founder of the Energy Probe Research Foundation Laurence Solomon, a previous speaker at the Heartland International Conference on Climate Change (Chapter 2), was described as:

an advisor to Task Force on the Global Environment (the Global 2000 Report) in the late 1970s, and at the forefront of movements to privatize and deregulate energy systems, reform foreign aid, and convert free roads to toll roads. The Deniers, his 2008 book exposing the global warming fraud, became the #1 environmental best seller in both Canada and the U.S. (Heartland Institute, n.d.)

The 2008 text written by Solomon subtitled, "*The world-renowned scientists who stood up against global warming hysteria, political persecution, and fraud*" critiqued legislation proposed by governments, international governance strategies, and activists and policymakers in climate and environmental justice movements.

The consistent messaging strategies used by this culmination of think tanks in Canada reflected the same trends observed in US and European opposition. While the practices are in some cases like those in Europe and the USA, and there are interlocks between them, Canada's opposition has unique components among the pluriverse of this global countermovement opposition ultimately entangled to the historical role that fossil fuels have played in the Canadian political economy.

5.5 Fringe Groups and Public Relations: Responding to Harper's Government (2006–2012)

From the election of Conservative Prime Minister Stephen Harper in 2006, the leader linked his candidacy and tenure with economic prosperity from natural resource exploitation (Goldenberg, 2015). Harper led as a *modern-day* conservative between 2006 and 2015. His ventures into the policies of climate change included the regressive actions of withdrawing from the Kyoto Protocol in 2011 and framing Canada as an energy superpower and oil exporter (Dalby, 2016). In 2014, he claimed that regulating the oil and gas sector would be a problematic economic policy and instead support an increase oil and gas exploitation and limit regulations (e.g., McDiarmid, 2014). Moreover, as O'Manique (2017, p. 3) noted, Harper's government transformed Canada into an energy superpower, increasing fossil fuel extraction and "disproportionately favoring the interests of the Alberta tar sands" coupled with climate denialism.

During this time, think tanks continued to operate to promote the continued use of fossil fuels, and resist calls for implementing effective climate legislation. For example, the Macdonald Laurier Institute (MLI) founded in 2010, received funding from the fossil fuel sector including "Canadian Gas Association, Teck Resources Limited, the Mining Association of Canada, the Charles Koch Foundation and the First Nations LNG Alliance" (Corporate Mapping, n.d.-b, n.p.). Similarly, the Montreal Economic Institute (MEI) also started to lobby heavily in the interests of the petroleum industry (Anderson, 2014). Like the Fraser Institute, the MEI focused some of its attention on Canada's energy profile, centralizing the key role of fossil fuels historically and to the present (e.g., MEI, 2014). Consistently, the organization promoted op-eds or work by those in the organization, emphasizing the important role of fossil fuels in Canada's political economy. Additionally, they critiqued attempts to restrict this whether that be international governance strategies such as the Kyoto Protocol (MEI, 2005), critiqued climate scientists drawing attention to Climategate (e.g., Elgrably-Levy, 2009), and campaigned to develop further fossil fuel extraction (e.g., Guenette, 2010). For example, in an op-ed by Desrochers and Guenette, they would criticize environmental activists claiming, "There is a certain fringe of the environmentalist movement whose members have almost nothing good to say about their fellow men and women" (2015, n.p.).

This position mirrored the likes of the Cornwall Alliance (Chapter 2) approaching the issue of protecting the environment in line with human mastery over the earth. Additionally, the organization's promotion of fossil fuel solutionism aligned with its emphasis on market-based solutions and preventive restrictions on the oil and gas industry. Moreover, the organization would position the Paris Agreement as a moment when the actions of non-Annex 1 countries would essentially have to step up to curb GHG emissions.

Accompanying think tanks, a small number of advocacy groups operated in the CCCM in Canada to elevate the messaging of counter-movement opposition and engage in a further PR campaign. These more fringe-like groups, particularly since the 2000s, garnered media attention with climate contrarians featuring heavily. Some of these individuals overlap with CCCM organizations in and beyond Canada. For instance, Ross McKitrick, whose work is cited under the Fraser Institute (e.g., Green, 2014a, 2014b), received several citations by Friends of Science (FOS) and appeared as "guest speaker" at the 11th Annual Friends of Science Luncheon with his paper titled *The Pause in Global Warming: Climate Policy Implications* (McKitrick, 2014). The promotion of uncertainty was central to the presentation emphasizing:

> *There is a prima facie case that the social cost of carbon has been overstated...The uncertainties will largely be resolved in the next 2–4 years... There is no downside to waiting for this, and considerable upside.* (n.p.)

The FOS formed in 2002 and when it emerged, claimed it wanted: "to challenge the questionable science and destructive economic impacts inherent in the politically inspired Kyoto Protocol" (n.p.). Much like the GWPF (UK), FOS was an advocacy organization focusing on climate change and environmental policy. Despite the growing scientific evidence, the organization consistently adopted various delaying discourses and did not waver from its position of denying human impact on the climate in some variation. In its earliest statement FOS claimed:

> *The Mission of the Friends of Science is to provide balanced, objective scientific knowledge to society in such a way that members of the public can make informed decisions on matters of concern to them...The need for responsible treatment of 'climate science' has arisen because 'climate science' entered into 'politics', long before sufficient factual data and hypotheses had been generally accepted.* (FOS, 2002)

One of FOS former directors Dr. Douglas Leahey was a figurehead for propagating delaying discourses into Canadian and USA media, and the political sphere. For example, in 2006 Leahey was cited in a letter to then Canadian Prime Minister Stephan Harper along with several other regular individuals operating among different organizations in the CCCM including Richard Lindzen, Robert Carter, and Patrick Michaels who had challenged the scientific consensus (Clark et al., 2006). The general premise of the 2006 letter critiqued the Kyoto Protocol and climate models as "alarmist" (2006, n.p.)." While in 2014 Len Meier (then president of FOS) claimed.

> *Friends of Science are absolutely not political activists nor a religious group. We were formed by a group of scientists 12 years ago because of the IPCC's abuse of science. Our mandate is to provide education concerning the science behind global warming.*

While adopting fringe-like positions disputing the scientific evidence on climate change, the FOS had been a central arm of Canada's fossil industry, receiving substantial corporate funding and close ties to the Alberta oil patch (Greenberg et al., 2011). The FOS was important in the Canadian context for promoting these alternative positions by running a vast PR campaign to influence the media, public perceptions and politics (Greenberg et al., 2011). These included producing promotional material such as the video *Climate Catastrophe Video* and radio ads. They highlighted the complex funding practice for FOS, reporting that funding initially went to the *Calgary Foundation* which had a specific funding division the *Science Education Fund*. From here, the funds would be distributed to organizations such as FOS, much like Donors Capital and Donors Trust in the US.

A similar fringe-type group, funded by Murray Energy and Robert E. Murphy was the International Climate Science Coalition (ICSC) which inaccurately claimed that the climate is *"always changing in accordance with natural causes and recent changes are not unusual"* (n.d., n.p.). Tom Harris, an Associate of APCO worldwide, created the organization. Harris and Evan Zelikovitiz coordinated its opening conference event, both of whom at the time, worked for APCO Worldwide (Climate Search, 2002), and who helped established The Advancement of Sound Science Coalition (ASSC) which was originally formed as an arm of the tobacco lobby (Climate Investigations, n.d.; Oreskes and Conway, 2010). Murray Energy also funded US CCCM organizations, including the CO_2 Coalition (Chapter 2). This CO_2 Coalition conceded the existence of CO_2 in the atmosphere was growing but says there is "no compelling reason" to link that to temperature rise (Waldman, 2019). Overtime, the ISCS continued to challenge the scientific consensus and was an offset launched after the New Zealand Climate Science Coalition (NZCSC) (Chapter 4) while connected to the Australian Climate Coalition (ASC) and the Climate Science Coalition of America.

I should also point out that Harris was a previous director of the Natural Resource Stewardship Project (NRSP). The NRSP launched in 2006 and aimed one of priority projects: "Understanding Climate Change…to counter the Kyoto Protocol and other greenhouse gas reduction schemes while promoting sensible climate change policy" (NRSP, 2006, n.p.). Thus, the organization was ultimately set up in a strategic response in a political opportunity to challenge the Kyoto Protocol. This organization emerged—only briefly—with the purposeful strategy of communicating the critiques of international governance institutions and environmental movements and centralizing individual responsibility. Importantly, while not critical of climate science, this narrative focused on individualizing, diverting responsibility away from the fossil industry to that of the public.

Continuing the trend of legitimizing fossil fuel use during a period where the government favored enhancing oil extraction, the media took on a broader role. Columnist and talk show host Ezra Levants *Ethical Oil: The Case for Canada's Oil Sands*, suggested that further exploitation is an essential social good echoing and giving further support for the Harper government's plans. As Aronczyk (2014) outlines, however, this allowed further propagation of the opposition positions, consistent with the overall aims of the C CCM. The book and subsequently the advocacy group associated with the text sought to expand their opposition by criticizing the aims of environmental activists (e.g., Ethical Oil, 2014a). At its core then, Ethical Oil would disseminate discourses of delay by positioning Canadian oil extraction as patriotic, tapping into Petro-nationalism and finding its way into mainstream Canadian media. For example, they criticized and attacked environmental groups and supportive foundatoins such as Environmental Defense, and Tides Canada (Ethical Oil, 2014b). Furthermore, they compared emissions from oil sands as minimal comparative to China, Taiwan, and the USA claiming that "an ethical country like Canada, we obviously take the environment a lot more seriously than the Chinese regime does" (Ethical Oil, 2011, n.p.). Such claims and propaganda not only appealed at a local level to those directly benefiting in Alberta but more to the broader Canadian public who would receive the economic fruits and stability from expanding these extractive practices.

5.6 Industry Groups, Extractive Populism and the Paris Agreement

During Harper's government and following the election of Justin Trudeau it became clear that Canada was not on track to meet any of its climate or emissions reduction targets. Nor did there appear the willpower to do so in the political realm, especially given the trend to legitimize and expand national oil exploitation. One reason for this was the rise in extractive populism which played a clear role in continuing the hegemony of the fossil fuel industry. Extractive populism can best be described as the extractive sector and constituted a core of the Canadian political economy and essential public good (Carroll et al., 2022). Thus, there was an appetite for increased fossil fuel extraction in the political realm and fostering in the minds of the public despite the need to reduce emissions and transition away from fossil fuels.

Industry associations representing the petroleum industry and advocating on behalf of its members, disseminating research, and working with government agencies to promote the industry's interests historically played a role in shaping climate policy and delaying climate action. They continued their campaign after the rise of Trudeau. As Carroll et al. (2021) outlined, these industry groups played a vital role in lobbying, conveying to their members that the implications of restrictions on their industries could impact the workers' economic security. This continued during Harper's government given his loyalties for the fossil fuel industry. Subsequently, as significant progress and positions were outlined leading up to the 2015 Paris Accord, these industry associations would target their campaigns to help ensure Canada remained a first-world Petro-state. Additionally, the geographic landscape of fossil fuel extraction in Canada presented different locations where extractive practices in different provinces could generate more income. The most resourceful of which was The Alberta Tar sands, parts of Saskatchewan, and British Colombia. Seeking to expand extractive practices, CCCM organizations made targeted campaigns in particular geographic locations as well as at a national level over time.

For example, the Canadian Association of Petroleum Producers (CAPP) was the largest and main lobbying organization for the fossil fuel industry in Canada (Environmental Defense, 2019). CAPP did not deny the evidence of climate change yet made significant attempts to undermine climate legislation that would reduce the role of the fossil fuel sector. Importantly, Graham et al. (2017) noted that CAPP held significant lobbying contacts, particularly, during the Harper administration seeking to impact relevant climate legislation and climate action plans. Between 2008 and 2015 alone, CAPP was the biggest donor to British Colombia political parties with only Imperial Oil totaling more. CAPP's lobbying would target politicians including William Bennett member of the British Colombia Liberal Party illustrating that political financing would cut across different political parties. While in the US finance was often concentrated very much in the Republican party, Canada's political economy would mean finance would cut across the political spectrum. When Tim McMillan took over as leader of the organization in 2014 it sparked a trend where CAPP would reinforce the fossil fuel industry's role in the economy. For example, in a speech to the Calgary Chamber of Commerce in 2014, McMillan spoke to a growing sense of nationalism and patriotism to counter the industry's criticisms. A previous member of the Government of Saskatchewa including Minister of Energy and Resources, McMillan would attempt to cultivate further support to expand the extractive industries following a significant decline in oil prices leading up to 2014, and the important role that fossil fuels could play in regrowing the economy.

Under new leadership, in 2014 CAPP launched the front group Canada's Energy Citizens (CEC). On its emergence, the CEC clearly articulated why fossil gas is important for the economy and essentially acted as a recruitment drive to garner supporters for the fossil industry, much like other advocacy and citizen groups such as the AEF (Chapter 4). In their initial discussion CEC promoted arguments that included the economic advantages of fossil fuel extraction, the energy security it provides, and existing regulatory protection mechanisms in place would minimize harm (CEC, 2014). For instance, they highlighted that "Fun Fact Friday: Canadian oil sands currently provide 514,000 jobs to Canadians. By 2028, that's expected to peak at 802,000 jobs" (CEC, 2014, n.p.).

The approach taken by the CEC and other groups in the Canadian arm of the CCCM reflected what Carroll et al. (2022) described as the new form of denialism. However, unlike the positions articulated by CCCM organizations when they initially emerged in the US, this new form of denialism would not necessarily deny the science of climate change. Instead, the focus was "manufacturing confusion" (p. 223) about the nature and extent of action and policies to respond to the climate crisis, advocating for voluntary and market-based approaches to GHG emissions reductions, and reinforcing to the public that actions taken are not necessarily a threat rather acceptable and on track to address the climate crisis. Moreover, the CEC was simply one example of the cultivation of a successful campaign to reinforce the importance of fossil fuels in Canada's political economy (Neubauer & Graham, 2021).

5.7 Conclusion

Countermovement opposition in Canada certainly has a long history. Active campaigns began around the late 1990s accelerating during the 2000s much like their American neighbors. Critically, the opposition positions articulated by countermovement organizations were deeply intertwined with the political economy of Canada. More specifically, as a first-world Petro-state, preventing restrictions on extractive industries would be of primary concern because it (1) would challenge the dominant economic growth patterns of the country and (2) would require a complete transformation of Canadian socio-cultural identity. Manifesting in think tanks, advocacy groups, and PR campaigns, these countermovement organizations would be able to reposition climate change and climate change science. It became something fraught with unknowns, could be framed as purely an economic issue, and would be used as a political issue that could be utilized by parties on both sides of the political spectrum.

References

American Enterprise Institute. (n.d.). Kenneth P Green. https://www.aei.org/profile/kenneth-p-green/. Accessed 17 September 2022.

Anderson, M. (2014, June 13). Look who financial post found to defend Canada's gifts to big oil. *The Tyee*. https://thetyee.ca/Mediacheck/2014/06/13/Financial-Post-Defends-Big-Oil/. Accessed 17 September 2022.

Aronczyk, M. (2014). Ethical oil: The case for Canada's oil sands. *Environmental Communication, 8*(1), 131–133.

Block, W. E. (1990). *Economics and the environment a reconciliation*. The Fraser Institute. Independent Institute. n.d. About us. https://www.independent.org/aboutus/person_detail.asp?id=318. Accessed 17 September 2022.

Canada Energy Citizens. (2014). Fuel for thought. economic benefits across Canada. https://web.archive.org/web/20150929080059/http:/

www.energycitizens.ca/get-the-facts/economic-benefits. Accessed 7 October 2022.

Canadian Energy Pipeline Association. (2003). Annual climate change workshop of the Canadian Energy Pipeline Association: Energy efficiency and greenhouse gas reduction opportunities, Calgary, AB (Canada), 23 September 2002. http://www.cepa.com/archives.htm. Accessed 17 September 2022.

Carroll, W. K. (2020). Fossil capitalism, climate capitalism, energy democracy: The struggle for hegemony in an era of climate crisis. *Socialist Studies, 14*(1). https://doi.org/10.18740/ss27275

Carroll, W. K., & Huijzer, M. J. (2018). *Who owns Canada's fossil-fuel sector?* Canadian Center for Policy Alternatives.

Carroll, W. K., Daub, S., & Gunster, S. (2022). Regime of Obstruction: Fossil capitalism and the many facets of climate denial in Canada. In *Handbook of anti-environmentalism* (pp. 216–233). Edward Elgar Publishing.

Carroll, W. K., Graham, N., Lang, M. K., Yunker, Z., & McCartney, K. D. (2018). The corporate elite and the architecture of climate change denial: A network analysis of carbon capital's reach into civil society. *Canadian Review of Sociology, 55*(3), 425–450. https://doi.org/10.1111/cars.12211

CGX Energy Inc. (2003). 2003 annual report. http://web.archive.org/web/20070203031545/http:/cgxenergy.ca/investors/CGX_AR03_part2.pdf. Accessed 17 September 2022.

Clark et al. (2006, April 6). In case you missed it...Open Kyoto to debate, sixty scientists call on Harper to revisit the science of global warming. *The Financial Post. Environmental and Public Work Committee*. https://www.epw.senate.gov/public/index.cfm/press-releases-all?ID=1E639422-7094-4972-83AF-EE40EE302D41. Accessed 17 September 2022

Climate Audit. (n.d.). Homepage. https://climateaudit.org. Accessed 17 September 2022.

Climate Investigations. (n.d.). APCO worldwide PR. https://climateinvestigations.org/public_relations-industry-climate-survey/apco_worldwide_pr/. Accessed 17 September 2022.

Climate Investigations. (n.d.). Imperial oil document trove. https://climateinvestigations.org/imperial-oil-document-trove/. Accessed 17 September 2022.

Climate Change Debate. (n.d.). Homepage. http://www.climatechangedebate.org/. Accessed 19 September 2022.

Climate Search. (2002). Kyoto's fatal flaws revealed Nov 12, 2002. http://web.archive.org/web/20030207194924/www.climatesearch.com/newsDetail.cfm?nwsId=54. Accessed 19 September 2022.

Coan, T. G., Boussalis, C., Cook, J., & Nanko, M. O. (2021). Computer-assisted classification of contrarian claims about climate change. *Science and Reports, 11*, 22320. https://doi.org/10.1038/s41598-021-01714-4

Cochrane, C. (2010). Left/right ideology and Canadian politics. *Canadian Journal of Political Science / Revue Canadienne De Science Politique, 43*(3), 583–605.

Conservative Transparency. (n.d.). Fraser Institute. Recipients. http://conservativetransparency.org/recipient/fraser-institute/. Accessed 17 September 2022.

Corporate Mapping. (n.d.-a). Canadian Energy Pipeline Association. https://www.corporatemapping.ca/profiles/canadian-energy-pipeline-association/. Accessed 17 September 2022.

Corporate Mapping (n.d.-b). Macdonald-Laurier Institute. https://www.corporatemapping.ca/profiles/macdonald-laurier-institute-2/. Accessed 6 September 2022.

Dalby, S. (2016). Geopolitics, ecology and Stephen Harper's reinvention of Canada. In *Handbook on sustainability transition and sustainable peace* (pp. 493–504). Springer.

Desrochers, P., & Guenette, J. (2015). Resourceful earth day (celebrate freedom, innovation). https://www.iedm.org/53014-resourceful-earth-day-celebrate-freedom-innovation//. Accessed 17 September 2022.

Doucet, J. A. (2004, May). *The Kyoto conundrum: Why abandoning the protocol's targets in Facour of a more sustainable plan may be best for Canada and the world* (p. 83). CD Howe Institute Backgrounder. https://www.cdhowe.org/public-policy-research/kyoto-conundrum-why-abandoning-protocols-targets-favor-more-credible-plan-may-be-better-canada-and. Accessed 17 September 2022.

Eaton, E. M., & Day, N. A. (2020). Petro-pedagogy: Fossil fuel interests and the obstruction of climate justice in public education. *Environmental Education Research, 26*(4), 457–473. https://doi.org/10.1080/13504622.2019.1650164

Ebell, M., & Smith, F. S. (2002, June 7). Joint letter to President Bush on the EPA's climate action report. https://web.archive.org/web/20020830110728/https://cei.org/gencon/027,03051.cfm. Accessed 18 September 2022.

Elgrably-Levy, N. (2009). Green religion. https://www.iedm.org/2180-la-religion-verte

Enserink, M. (2004). Moscow meeting bogged down in acrimony. *Science, 305*(5682), 319. https://link.gale.com/apps/doc/A119951441/AONE?u=googlescholar&sid=googleScholar&xid=8f64f32c. Accessed 18 September 2022.

Environmental Defense. (2019, October). The single biggest barrier to climate action in Canada: The oil and gas lobby. https://environmentaldefence.ca/wp-content/uploads/2019/10/EDC-Oil-Climate-Action-Oct-2019.pdf. Accessed 18 September 2022.

Ethical Oil. (2011). Mythbusting: Are the Oilsands major greenhouse gas emitters. https://web.archive.org/web/20140929120930/http://www.eth

icaloil.org/news/myth-busting-are-greenhouse-gas-emissions-from-the-oil sands-ruining-the-atmosphere/. Accessed 7 December 2022.

Ethical Oil. (2014a). Ethical Oil. https://web.archive.org/web/2014a0904172617/http://www.ethicaloil.org/. Accessed 7 December 2022.

Ethical Oil. (2014b). Press Release. https://web.archive.org/web/2014b0623202757/http://www.ethicaloil.org/news/press-release/. Accessed 7 December 2022.

Fast, T. (2014). Stapled to the front door: Neoliberal extractivism in Canada. *Studies in Political Economy, 94*, 31–60.

Fraser Institute. (n.d.-a). Environment. *Fraser Forum*. https://www.fraserinstitute.org/blogs/category/environment. Accessed 18 September 2022.

Fraser Institute. (n.d.-b). Environment. https://www.fraserinstitute.org/blogs/category/environment. Accessed 17 September 2022.

Fraser Institute. (2002a). Environment. https://web.archive.org/web/2002a0604183910/http://fraserinstitute.org/environment/index.asp?snav=en. Accessed 18 September 2022.

Fraser Institute. (2002b). Taken by storm: The troubled science, policy and politics of global warming. https://web.archive.org/web/20030120180917/http:/www.fraserinstitute.org/shared/readmore.asp?snav=ev&id=109. Accessed 17 September 2022.

Fraser Institute. (2002c). *Challenging perceptions: Twenty-five years of influential ideas: 1974–1999. A retrospective*. The Fraser Institute.

Friends of Science. (n.d.). About us. https://friendsofscience.org/about/about-us.html. Accessed 17 September 2022.

Friends of Science. (2002). Our purpose. https://web.archive.org/web/20021018082749/http:/www.friendsofscience.org/our_purpose.htm. Accessed 17 September 2022.

Frontier Center for Public Policy. (n.d.). About. https://fcpp.org/about/. Accessed 18 September 2022.

Frontier Center for Public Policy. (2002). A skeptics guide to global warming. https://fcpp.org/2002/10/14/a-skeptics-guide-to-global-warming/. Accessed 17 September 2022.

Goldenberg, S. (2015, October 16). Canada election: How Stephen Harper's fossil fuel gamble may have backfired. *The Guardian*. https://www.theguardian.com/world/2015/oct/16/canada-election-stephen-harper-fossil-fuel-backfires. Accessed 17 September 2022.

Guenette, J. (2010). The Oilsands: A source of prosperity for all. https://www.iedm.org/33820-the-oilsands-a-source-of-prosperity-for-all/. Accessed 17 September 2022.

Gunster, S., Fleet, D., & Neubauer, R. (2021). Challenging petro-nationalism: Another Canada is possible? *Journal of Canadian Studies, 55*(1), 57–87. https://doi.org/10.3138/jcs.2019-0033

Graham, N., Daub, S., & Carroll, W. (2017). *Mapping political influence*. Canadian Center for Policy Alternatives.
Green, K. (2015, August 25). It's abundantly clear—Pipelines are safer than rail. https://www.fraserinstitute.org/blogs/its-abundantly-clear-pipelines-are-safer-than-rail. Accessed 17 September 2022.
Green, K. (2014a, November 6). Climate change: It's not the science, it's the policy. https://policycommons.net/artifacts/419535/climate-change/1389820/. Accessed 17 September 2022.
Green, K. (2014b). Kenneth Green debates climate change on CBC. https://www.fraserinstitute.org/blogs/watch-kenneth-green-debate-climate-change-on-cbc. Accessed 17 September 2022.
Green, K., & Jackson, T. (2015, October 6). Pipelines are safer, cheaper and greener. https://www.fraserinstitute.org/blogs/pipelines-are-safer-cheaper-and-greener. Accessed 17 September 2022.
Greenberg, J., Knight, G., & Westersund, E. (2011). Spinning climate change: Corporate and NGO public relations strategies in Canada and the United States. *International Communication Gazette, 73*(1–2), 65–82. https://doi.org/10.1177/1748048510386742
Heartland Institute. (n.d.-a). David Wojick. https://www.heartland.org/about-us/who-we-are/david-wojick-ph-d. Accessed 17 September 2022.
Heartland Institute. (n.d.-b). Lawrence Solomon. https://www.heartland.org/about-us/who-we-are/lawrence-solomon. Accessed 17 September 2022.
Heartland Institute. (n.d.-c). Richard Stroup. https://www.heartland.org/about-us/who-we-are/richard-stroup. Accessed 17 September 2022.
Holle, P. (1997). Lets be careful about global warming. *Commentary*. https://fcpp.org/1997/01/26/lets-be-careful-about-global-warming/. Accessed 17 September 2022.
Illarinov, A. (2004). The Kyoto Protocol: To ratify or not to ratify? https://web.archive.org/web/20040409105527/http://www.fraserinstitute.org/files/illarionov.pdf. Accessed 17 September 2022.
Jones, L. (1997). *Global warming: The science and the politics*. Vancouver. The Fraser Institute.
McDiarmid, M. (2014, December 9). Stephen Harper says oil and gas regulations now would be 'crazy'. *CBC News*. https://www.cbc.ca/news/politics/stephen-harper-says-oil-and-gas-regulations-now-would-be-crazy-1.2866306. Accessed 17 September 2022.
McKie, R. E. (2021). Obstruction, delay, and transnationalism: Examining the online climate change counter-movement. *Energy Research and Social Sciences*, 80.https://doi.org/10.1016/j.erss.2021.102217
McKitrick, R. (2014). The "pause" in global warming: Climate policy implications. https://friendsofscience.org/assets/documents/McKitrick2014_ThePause.pdf. Accessed 28 September 2022.

McKitrick, R., & Wigle, R. (2002, October). *The Kyoto Protocol: Canada's risky rush to judgement* (CD Howe Institute Commentary. No. 169). https://www.cdhowe.org/public-policy-research/kyoto-protocol-canadas-risky-rush-judgment. Accessed 28 September 2022.

McKitrick, R., D'Aleo, J., Khandekar, M., Kininmonth, W., Essex, C., Karlen, W, Karner, O., Clark, I., Murty, T., & O'Brien, J. J. (2007). *Independent summary for policymakers IPCC fourth assessment report*. Vancouver. The Fraser Institute.

Montreal Economic Institute. (2005). Montreal Economic Institute questions leaders on the economic impact of the Kyoto Protocol. https://www.iedm.org/3297-montreal-economic-institute-questions-leaders-on-the-economic-impact-of-the-kyoto-protocol/. Accessed 28 September 2022.

Montreal Economic Institute. (2014). Canadas energy profile in 40 questions. https://www.iedm.org/wp-content/uploads/2019/10/energy-profile-en.pdf. Accessed 17 September 2022.

Natural Resource Stewardship Project. (2006). Background. http://web.archive.org/web/20070323204851/http://www.nrsp.com/background.html. Accessed 28 September 2022.

Nikiforuk, A. (2010). *Tar sands: Dirty oil and the future of the continent*. Greystone Books.

Neubauer, R., & Graham, N. (2021). Fueling the subsidized public: Mapping the flow of extractivist content on Facebook. *Canadian Journal of Communication, 46*(4), 905–938.

Parson, S., & Ray, E. (2016). Sustainable colonization: Tar sands as resource colonialism. *Capitalism, Nature, Socialism, 29*(3), 68–86. https://doi.org/10.1080/10455752.2016.1268187

Piggot, G. (2018). The influence of social movements on policies that constrain fossil fuel supply. *Climate Policy, 18*(7), 942–954.

Schneider, N. (2006). Earth day—Why so glum? https://www.fraserinstitute.org/article/earth-day-why-so-glum?language=en. Accessed 18 September 2022.

Schwanen, D. (1997, October 28). Policy on climate change should aim to reduce risks, avoid unnecessary costs, says C.D. Howe Institute. Comminque. https://www.cdhowe.org/public-policy-research/confronting-greenhouse-challenge-matching-protection-risk. Accessed 28 October 2022.

Sharman, A. (2014). Mapping the climate skeptical blogosphere. *Global Environmental Change, 26*(May), 159–170. https://doi.org/10.1016/j.gloenvcha.2014.03.003

Smith, H. A. (2009). Political parties and Canadian climate change policy. *International Journal, 64*(1), 47–66.

Tynkkynen, N. (2010). A great ecological power in global climate policy? Framing climate change as a policy problem in Russian public discussion.

Environmental Politics, 19(2), 179–195. https://doi.org/10.1080/09644010903574459

Waldman, S. (2019, December 17). Murray Energy a major funder of climate denial groups. *E&E News. Green Wire.* https://www.eenews.net/articles/murray-energy-a-major-funder-of-climate-denial-groups/. Accessed 28 October 2022.

Whiteman, G., & Mamen, K. (2002). *Meaningful consultation and participation in the mining sector?: A review of the consultation and participation of indigenous peoples within the international mining sector.* North-South Institute.

Wrenn, C. (2022). Building a vegan feminist network in the professionalized digital age of Third Wave animal activism. In E. Cudworth, R. E. McKie & D. Turgoose (Eds.), *Feminst animal studies: Theories, practices, politics*. Routledge.

CHAPTER 6

Manifestations of the Climate Change Counter Movement in the Global South

6.1 Understanding the CCCM in the Context of the Global South

In the previous chapters, I have explored the operations of the CCCM in countries in the Global North. One reason was that my research and existing literature had been concentrated in these locations. However, that is not to say CCCM actors did not operate outside these countries. As such, this chapter synthesizes some existing knowledge on the history of some CCCM organizations and actors in different countries within the Global South.

Before outlining these CCCM operations, it is essential to consider the various social, political, economic, and cultural contexts in which these organizations and actors emerged.

First, while nations in the Global South are not homogenous and variation exists within and between them, there are five consensus points. One, in the previous chapters, the discussion predominantly revolved around fossil fuel and related industries. More specifically, I articulated that fossil fuels have been central to the historical development of the current global political economy, from which no country is immune. The political economy perspective of unequal ecological exchange (e.g., Foster & Holleman, 2014; Jorgenson, 2006) can help us understand the impacts of climate change and why CCCM operations emerged in some parts of the Global South, particularly concerning transnational forces.

R. E. McKie, *The Climate Change Counter Movement*, https://doi.org/10.1007/978-3-031-33592-1_6

The perspective suggests "global structures and relationships that shape the unequal distribution of environmental harms and human development" (Givens et al., 2019, p. 2). In other words, capitalism has relied on the pillaging and theft of land, resources, and cultures from many countries in the Global South (Foster & Clark, 2004). Consequentially, we end up with the unequal distribution of environmental harms and subsequent impacts of climate change, whereby countries in the Global South are significantly more vulnerable (e.g., Bordner et al., 2020).

Secondly, while not directly associated with the theorization of unequal ecological exchange, we can identify how countries in the Global South play a role in international governance, particularly from the UN Conference in 1992, which is different from those in the Global North. Under the 1997 Kyoto Protocol, many countries in the Global South were placed under non-Annex I countries. Non-Annex I countries were mainly developing countries and the most vulnerable to the impacts of climate change. Moreover, some, although not all, of these countries relied heavily on income from fossil fuel production and polluting industries or extractive-based economies. Therefore, they were also more vulnerable to the economic impacts of responding to climate change that could put restrictions on these heavily polluting industries (UNFCCC, n.d.-b). While not under the same conditions as those in Annex I, non-annex I countries could also engage in GHG emission reduction projects (Lau et al., 2009), predominantly Clean Development Mechanisms and Joint Implementation programs. Therefore, before the 2015 Paris Agreement, international governance efforts focused on industrialized countries reducing their GHG emissions substantially while offering financial support for countries most vulnerable to the impacts of climate change.

Concerning countermovement opposition and broader climate obstruction, the global context around international emissions reductions and commitments could invoke a Polluter Pays principle (Feng & Buhi, 2010). The foundation of this principle is an economic production model whereby those most responsible should also pay the direct costs and ensure further harm prevention (Champéroux, 2019). This then applies a meaningful and supported argument to mitigate progressive GHG emissions reduction targets likely successful if countries that have most contributed to human-caused climate change fail to act meaningfully. On the one hand, the narratives from countermovement organizations may well be justified in that they present a possible legitimate position.

On the other, and as evidenced throughout this chapter, other tools and interactions between Global North based countermovement organizations and those in the Global South indicate familiar and mirrored opposition positions that were all too characteristic of countermovement forces.

Thirdly, it is questionable whether a countermovement characterized as part of a response to challenging the DSP can fully explain or help understand how such a movement could apply to the Global South. In other words, the DSP is very much an American-specific worldview. That said, it certainly helps us understand more generally industrialized and neoliberal societies and human–nature relationships (Bogert et al., 2022). If therefore, industrialization and economic development went hand in hand with the quality of life synonymous with consumption, this DSP paradigm is relevant in the context of a global political economy (e.g., Fotopoulos, 2008). However, environmental concern and how people in the Global South understand their relationship with nature is different (e.g., Polonsky et al., 2014). Nevertheless, global development pathways are characterized by GDP growth as an indicator of quality of life. Therefore, where extractive industries associated with increasing GHG emissions became the central focus of economic development, arguments that would emerge from countermovement organizations would be able to draw on this position, particularly regarding reducing poverty and issues of social justice.

Fourthly, from the 1990s onwards, developing countries in the Global South saw exponential growth in homegrown think tanks (Datta & Young, 2011). Such valuable institutions enabled these countries to operate as policy developers, which governments could adopt. Think tanks served a strategic purpose, especially in the face of growing international agreements such as the UNFCCC and subsequent mechanisms that would ultimately impact domestic-level climate and related legislation. Across the Global South, some think tanks were non-profit making, government, or privately funded, often with different ideological purposes. However, similar to USA libertarian and neoliberal counterparts, these organizations did emerge and operate with specific issues in mind.

Moreover, they had ideological and party-political aims, making them a clear breeding ground for disseminating discourses of delay. Therefore, a network infrastructure emerged in some Global South countries to facilitate the spread of ideas and the growing influence of think tanks in the political realm. The infrastructure for mobilizing opposition efforts was then responsive to growing environmental justice movements and increasing engagement in national and international climate-related policy.

Finally, the impact of South-South cooperation programs of investments is essential to these discussions. Described as "the technical cooperation among developing countries…used by the states, international organizations, academics, civil society and the private sector to collaborate and share knowledge, skills and successful initiatives in specific areas etc." (UN, n.d., n.p.), South-South cooperation ultimately reflected a shift in world politics and economics seeking an alternative vision around global capitalist development, including the development of reorganized pathways which would reflect a liberating transformation. More specifically, growth in the Global South had been one of northern dominated political and economic development and legacies of colonial histories (Gray & Gills, 2016). South-South cooperation involves at a minimum, increased lending between countries in the Global South in contrast to the significant contributions of the Global North and corporations. While there are differences, South-South cooperation ensured development projects, including extractive projects, are concentrated between countries in the Global South.

6.2 CCCM Operations and the Mont Pelerin Society: 2000–2007

The notion of a spread of ideas provides an important starting point to examine how some opposition organizations came to operate in Global South. In Chapter 3, I explored and identified the importance of transferring knowledge and ideas to different countries in a policy network to support the interests of corporations and make neoliberal transformations across the economy. The clearest historical example is the Mont Pelerin Society (MPS), founded in 1947. Built by a group of economists, businessmen, and civil servants, they came together to forge an "intellectual" collective that would challenge the changes to the political economy, particularly the rise of collectivism following the end of World War II (Plehwe, 2015).

This group of intellectuals came from various countries with different backgrounds and experiences. As a result, it presented the opportunity to forge a global idea that could shape ideas of what a global political economy could look like. The group's purpose was to present and support a doctrine that could filter and lead to the development of an economy that ensured "privatization, deregulation, and financial and trade liberalization" (Plehwe, 2015, p. 8). For a more in-depth discussion on the MPS, see Mirowski and Plehwe (2015). Notably, the MPS had been central in constructing and advancing the fossil fuel political economy since the 1930s including leaders such as the Koch Brothers forming part of the organization (Walker, 2020). As such, the proposals of the MPS sought to follow a form of classical liberalism and remove barriers that would restrict the market's image and success intertwinned with fossil fuel extraction (Mirowski, 2014).

Given the transnational formation of the MPS, it is unsurprising that individuals would be able to facilitate the spread of delaying discourses into the Global South. For example, in 2006, the MPS held its general meeting in Guatemala (MPS, 2007) following previous regional meetings in 1973 and 1990. The 2006 meeting was organized by Giancarlo Ibarguen, who in 2011 became MPS secretary of the Board of Directors. Ibarguen was a Professor of Economics and former President at the Guatemalan Francisco Marroquin University (UFM) and himself disseminated delaying discourses.[1] In addition, the university was a site for several seminars between 2013 and 2014 involving Ibarguen and the Cato University (Cato Institute, n.d.). Ibarguen was also a former US-based Liberty Fund board of directors member, founded by Pierre F Goodrich in the 1960s. The Liberty Fund held conferences on the issue of the environment, including a 2012 event with a reading list that included the work of Indur Goklany and the GWPF (Chapter 3) (Liberty Fund, n.d.). Similarly, Christopher Lingle was an economics professor at UFM and an adjunct scholar at the CIS (Chapter 4). Writing for the US Foundation for Economic Education (FEE), in 2003, Lingle suggested that the science of climate change remains unsettled (see also Lingle, 2004).

Individuals promoting discourses of delay outside the USA and Europe were important for spreading ideas and globalizing countermovement opposition. The interlocks between MPS members and think tanks filtered a consistent message on the issue of climate change in an attempt to influence policies, particularly in Latin America. One example was the Instituto Liberdade (Liberty Institute), a Brazilian think tank originally

[1] The university's associated foundation received funding from the Pierre Foundation.

founded in 1986 which prioritized "research, production and dissemination of educational and cultural goods... based on the principles of individual rights, limited and representative government, respect for private property, contracts and free enterprise" (Instituto Liberdade, n.d.).

Between 2004 and 2017, the organization hosted an activities program and created a report repository on climate change and related environmental issues. In addition, it co-sponsored several Heartland Institute's International Conference on Climate Change (Chapter 2) and promoted the work of climate contrarians connected to other CCCM organizations, particularly from the USA, Canada, and Europe (see Instituto Liberdade, 2010). While we cannot determine that funding from foundations was directed to climate-related content, conservative donations to the organization exist, much like the USA element of organized obstruction. For instance, between 2005 and 2011, the organization received $95,000 from Pierre F. and Enid Goodrich Foundation. The foundation also funded other countermovement organizations, including the Heritage Foundation, Cato Institute, and the CEI (Conservative Transparency, n.d.). The institute also formed part of the Fraser Institute's Economic Freedom and the World Global Network. Within this network were other think tanks in the Global South that had promoted discourses of delay including the Nassau Institute (Bahamas), Center for Civil Society (India), and the Institute for Democracy and Economic Affairs (Malaysia). See Fraser Institute (Economic Freedom of the World, n.d.) for a list of organizations.

Several Instituto Liberdade academic councils and previous boards of directors, including Henri Siegert Chazan, Margaret Tse, and Leônidas Zelmanovitz, had earlier ties to the MPS (see De Smog, n.d.). Therefore, the advocation by MPS members to shape political and economic institutions in Latin America was no surprise. In fact, it was framed as an *enlightenment* project in a move towards a capitalist-based economy and avoiding the rise in socialism and/or welfarism, which had previously swept across North America and Europe (Liggio, 1990). As such, the ideas of MPS would have to be transferred into every policy issue, including climate change. The result then was a network created that helped diffuse framing on the issue of climate change—often ideologically aligned—that could be streamed across the political process and implemented in national and global policy.

6.3 The Civil Society Coalition on Climate Change and Think Tank Networks (2007–2009)

Described in Chapter 2 was a coalition of organizations that would represent the fossil fuel industry. However, coalitions could also unite think tanks across countries. One such example that acted as a vehicle for diffusing US-grown countermovement arguments into a selection of Global South-based think tanks was the Civil Society Coalition on Climate Change (CSCCC). The CSCCC was founded in the USA and released its first report in 2007. Member organizations existed in several countries outside the Global North (see Table 6.1). Creating a coalition of organizations across countries was an effective tool deployed as part of a countermovement strategy like that in Europe to diffuse a consistent message and create a "policy" strategy that could produce consistent outcomes across multiple countries. As Plehwe (2014, p. 101) notes, "Such networks are designed to promote or disrupt political discourse." Therefore, this knowledge exchange would prove helpful for (1) expanding and promoting variations of neoliberalism that could be weaved into policy in every country and (2) disseminating similar discourses of delay into policy in every country.

The CSCCC's 2007 opening report commentated on the lack of and misrepresentation of evidence on climate change. It employed discourses of delay on the issues of social justice, and economic growth, while proposing free market approaches could best address environmental problems. Authors of the report included Paul Reiter,[2] Indur Goklany, and Douglas Southgate, and the report concluded:

> *The Civil Society Report on Climate Change concludes that such emissions caps [proposed in a post-Kyoto treaty] would be counterproductive: they would undermine economic development, harm the poor, and would be unlikely to address the problem of climate change in a meaningful way.* (2007)

Within the report, the authors raised arguments on the benefits of higher CO_2 levels because they could replenish forest decline in the Amazonian region. The authors emphasized that increasing CO_2 could improve and

[2] Paul Reiter appeared in the documentary the Great Global Warming Swindle along with other climate contrarians. Citation. https://www.sourcewatch.org/index.php?title=The_Great_Global_Warming_Swindle. Accessed 29 November 2022.

Table 6.1 The civil society coalition on climate change

Organization name (Country)	
Alternate Solutions Institute (Pakistan)	Initiative for Public Policy Analysis (IPPA) (Nigeria)
Alabama Policy Institute (USA)	Institute for Liberty and Analysis of Policy in
Asociación de Consumidores Libres, (Costa Rica)	Government (INLAP) (Costa Rica)
Association of Liberal Thinking (Turkey)	Beacon Center Tennessee (USA)
Bluegrass Institute for Public Policy (USA)	Institut Constant de Rebecque (Switzerland)
CGC Forum (China)	Institute for Free Enterprise (Germany)
Cathay Institute of Public Affairs (China)	Institute for Market Economics (Bulgaria)
CEDICE (Venezuela)	Institute for Economic Analysis (Russia)
Center for Human Development (Centro de Innovación y Desarrollo Humano) (Uruguay)	Institute for Public Affairs (Australia)
Center for Political Studies (CEPOS) (Denmark)	Instituto de Libre Empresa (Peru)
China Sustainable Development Research Center, Capital University of Business and Economics (China)	Instituto Liberdade (Brazil)
CEPPRO (Paraguay)	International Policy Network (UK)
CIIMA-ESEADE (Argentina)	Instituto Bruno Leoni (Italy)
CORE (USA)	Jerusalem Institute for Market Studies (Israel)
European Center for International Political Economy (European Center for Economic Growth) (Austria)	John Locke Foundation (USA)
Fraser Institute (Canada)	Libreral Institut (Switzerland)
Free Market Foundation (South Africa)	Liberalni Institute (Czech Republic)
Frontier Center for Public Policy (Canada)	Liberty Institute (India)
Fundacion Atlas 1853 (Argentina)	Lion Rock Institute (Hong Kong)
Fundacion Libertad (Argentina)	Lithuanian Free Market Institute (Lithuania)
Ecuadorian Institute of Political Economy (Ecuador)	Minimal Government Thinkers (The Philippines)
Hayek Institute (Austria)	New Economic School (Georgia)

(continued)

Table 6.1 (continued)

Organization name (Country)	
CEES (Guatemala)	New Zealand Business Roundtable (New Zealand)
Center for Ethics and Technological Development (Nigeria)	IMANI (Ghana)
Center for Economic Transition (China)	Adam Smith Institute (UK)
Institute for the Study of Humane Action (Peru)	Americans for Tax Reform (USA)
Libertad y Desarrollo (Chile)	Cascade Policy Institute
Reason Foundation	Center of Human Affairs (Burkino Faso)
The Taxpayer's Alliance (UK)	Institute for Democracy and Economic Affairs (Malaysia)
	Democracy and Market Institute (Chile)
	Washington Policy Center (USA)

expand global timber production. By strengthening markets with little to no government intervention it would help address climate change, advance development, prevent rising poverty, and reduce the need for foreign aid from developed countries. While much of the report was aimed at US readers, a chapter by Southgate and Sohngen on agribusiness could have resonated with readers where land use changes to expand the agribusiness sector. This was an important part of developmental plans in some of these countries.

Demonstrating support for the deregulation of the agribusiness sector would become the focus of CCCM organizations in some countries in the Global South and broader political and corporate activities to obstruct efforts to reduce emissions. One example is Brazil, where the agribusiness sector and supporters challenged climate-related regulation (e.g., Milani & Chavez, 2022). This reflected the biodiversity and land within the country and the role played by the agribusiness sectors in stimulating economic growth as part of the developmental strategy. Notably, from 2004, emissions from agriculture made the largest contribution to GHG emissions in Brazil (European Parliament, 2022). In addition, it reflected the racialized history of land rights, and land ownership that have been embroiled and entangled in the weakening of environmental legislation to expand opportunities for the "Appropriation" of land, particularly from indigenous communities, without compensation and recognition. Viola and Franchini (2014) documented that the agribusiness sector tended to see climate and environmental-related policy expansion as a threat to industry expansion. And even with illegal deforestation (e.g., Coelho-Junior et al., 2022), the existing regulations only exacerbated deforestation. Rajão et al. (2020) noted that only a common solution with international shared responsibility could balance development with GHG emissions reductions. Nevertheless, we could say that this is when opportunities arose to disseminate discourses of delay that would reflect the social, political and economic makeup of some countries in the Global South. Therefore, countermovement organizations such as the Instituto Liberdade used the CSCCC outputs to support a series of oppositional positions.

A think tank network of CCCM connected by their affiliation to the CSCCC and ideologically aligned, successfully campaigned and mobilized to promote various neoliberal and libertarian philosophical approaches into politics. It was part of a project to expand these ideas so that they could be implemented into policy. This included facilitating opportunities for increased privatization of resources, commodifying and concentrating the wealth outside the state, and challenging greater state control over natural resources and commodities. In addition, with ties to contrarians, these networks penetrated the discourse of some countermovement organizations in the Global South.

One reason for activating this think tank network was to align responses and critiques of climate change science and policy with the move to promote a particular economic policy in developing nations: the Washington Consensus (e.g., Williamson, 1990). The Washington Consensus was the dominant discourse used since the late 1980s to understand the development and a proposed development strategy in the Global South prescribed by large financial and governance institutions. While a complete discussion on the consensus is beyond the purview of this text, in short, the consensus is a paradigm through which the adoption of policies based on macroeconomic stability, open economies, and increased privatization and deregulation would be stabilizing forces to generate growth (e.g., Gore, 2000). In practice, this paradigm was facilitated and implemented by the International Monetary Fund (IMF) and World Bank, among others such as nation-states attempting to configure a global and consistent development approach in developing nations.

Although this was not a match made in heaven. Neoliberal trends and its connections with the Washington Consensus would result in an apparent growth in these nations, but it also rose in tandem with the significant rise and success of think tanks in the region of Latin America corresponded with the rise of the Pink Tide. During the late 1990s and early 2000 across the region, the Pink Tide encompassed a political shift where governments sought to reject the Washington Consensus and proposed and implemented a form of developmental model in some way that was distinct. For a further discussion on the Pink Tide, see Saputra (2019). It is important to note that this theoretical model has different interpretations and critiques on how far it can explain development in the Global South. However, what it did do was reflect the ideological trends that flow through the arguments of countermovement organizations that spread discourses of delay. That is, the issue of climate change was framed as a free market issue associated with neoliberal reforms proposed by these think tanks.

For example, Andes Libres was a Peruvian think tank promoting the values of individual freedom, limited government, and a free market economy. As for its role as a CCCM, while not clearly outlining its position on climate change, it cited the work of contrarians including Donald Boudreaux and the text *The new global enemy of the Poor: Environmentalists* (2006) and promoted ideas from different countermovement organizations and contrarians, particularly in the USA (e.g., Goklany, 2008; Michaels, 2010). Following the Austrian School of Economics, much like the Guatemalan connections described above, staff in the organization included Alex Chaufen (also an MPS member), Lawrence Reed (President of the FEE) and Paul Driessen, a senior member of the USA based Center for the Defense of Free Enterprise.[3] The criticisms mirrored countermovement arguments that manifested in the USA and Europe and transcended national boundaries.

Reaching other areas of the Global South, the CSCCC messages and the ongoing ties to this think tank network permitted the diffusion of delaying discourses to reach new people and policymakers. Founded in 1995, the Nassau Institute in the small island state of

[3] The Center for the defense of free enterprise was one of the founding organizations of the CCCM. Citation. Climate Investigations (n.d.) Competitive Enterprise Institute NYT Ad Signatories Got $10 Million from Exxon. Davies (2016) https://climateinvestigations.org/nyt_ad_signatories_got_10_million_from_exxon/. Accessed 1 Dec 2022.

The Bahamas promoted the value of free market economies, economic growth, limited government, and private property rights (Nassau Institute, n.d.). As a small island state, vulnerable to the impacts of climate change, the Bahamian government first ratified the Kyoto Protocol in 1999, later signing the 2015 Paris Agreement. While the country itself and the government ultimately recognized the need for significant action to address the climate crisis through involvement and quick ratification of the Kyoto Protocol (e.g., UNFCCC, n.d.-a). Despite the overwhelming evidence of ecological harm, an attempt although not necessarily successful, countermovement opposition emerged.

Instead, the think tank would target policy by promoting the spread of ideas, and a campaign intended to undermine and delay the implementation of government policy. To do this, not only would the institute become part of the CSCCC, but it would also promote the work of contrarians that sit within the CCCM and connect organizations across countries (e.g., Nassau Institute, 2002), and forge its own positions by connecting the Bahamian experience with free market approaches to climate change resolutions. For instance, in response to the ratification of the Kyoto Protocol, a post by the institute outlined at the International Expert's Forum that, "Signing on to international protocols on Global Warming, makes little sense for the Bahamas…particularly when we consider what our tiny archipelago's carbon dioxide emissions total" (2003, n.p.). Furthermore, the post suggested that "The IEF's response to the end of the world hysteria was to inform Bahamians that there are scientists whose research does not support that of those quoted by Mr. Weech, the Director of the National Climate Change Commission (NCCC)" (n.p.). Referencing other CCCMs, including the Heartland Institute (e.g., Nassau Institute, 2004), continuing the trend of think tank activities in the US and Europe, the organization wanted to add this delaying discourse into the mainstream. Over time they continued presenting these discourses of delay, referring to Climategate incorporating positionality Bjorn Lomborg to justify their position (Nassau Institute, 2009). While the organization did not explicitly deny the science of climate change, it critiqued scientists and ultimately used a common delaying technique observed across the CCCM.

Similarly, the Nigerian Initiative for Public Policy Analysis (IPPA), a member of the CSCCC, would frame climate change as an issue only resolved through market-based mechanisms. Identifying as a leading Nigerian research institute promoting a "free and open society" (n.d., n.p.), from its inception, the organization centralized market-based approaches to addressing environmental problems with an emphasis on property rights as the best mechanisms for protecting the environment and conserving natural resources (IPPA, 2004). In 2009, the director of the IPPA, Thompson Ayodele, described the outcomes of the Copenhagen Agreement and other international interventions as follows:

> *Environmental groups from rich countries have for years waged a campaign against those in poor countries who want to harness their natural resources for economic growth. Their efforts threaten to do lasting harm to the aspirations of millions of poor people in Africa, Asia and Latin America, and must be resisted at all times and in all places.*

Materially, oil extraction was presented as crucial component of and underpinned Nigeria's economic development, supported and facilitated by state–corporate relationships invested in expanding oil extraction in the nation (e.g., Manby, 1999). This close state–corporate relationship was evidenced by regimes of rent-seeking and the accumulation of capital from fossil fuel extraction from a faction of the country's elites that optimized their relationships with fossil fuel multinational corporations (e.g., Omeje, 2017). Moreover, during the 1990s and early 2000 the role of petroleum companies such as Shell engaged in lobbying efforts in Washington, DC to ensure they could help craft Nigeria's governmental reforms (Kalu, 2006). As such, infrastructure would be in place to harness the centrality of market-based mechanisms to support and expand Nigeria's extractive economy.

This messaging associated with increased oil extraction would become optimal for this think tank particularly because oil extraction was framed as an incentive for development. Historically it was combined with notions of sovereignty, the right to develop, and protecting human rights (e.g., Doyle & Lockhart, 2012; Kolk & Levy, 2001; Manby, 1999). Broken down, Eberlein (2006) described the complexity of oil exploitation particularly in the Niger Delta region. They highlighted the complex multifaceted organizational and institutional actors in the capture and management of extraction. These included investments and corporate development from multinational corporations, "development grants from bi and multilateral donors, election observer missions and reports by human rights groups..." (p. 578). In turn, disseminating discourses of delay that adopted arguments of sovereignty, the right to develop, and reasserting ownership of resources, with economic growth meant these messages could be more receptive by policymakers and public opinion.

The spread of neoliberalism and framing strategy on the issue of climate change through the vehicle of the CSCCC reached Asia. In the Philippines, Minimal Government Thinkers, once a co-sponsor to the 2009 Heartland Institute Climate Change Conference, cemented its position as a countermovement organization. Nonoy Oplas was the organization's president, and in 2009, he reported on the scientific discussions on climate change, drawing on summaries from the conference (Oplas, 2009). Oplas would similarly report on "climate alarmism" and critiques of activists and collective action, drawing on papers from various news sources and blogs (Oplas, 2010). These messages followed a consistent pattern of neoliberal thought and, in this case, promote free market environmentalism. Likewise, in Turkey was the Association of Liberal Thinking. A think tank that promoted neoliberal thought would form its own Center for Environmental Studies (Association of Liberal Thinking, n.d.), again supporting the Heartland Institutes conference (see also McKie, 2021).

It is evident throughout this illustration that a think tank network disseminating discourses of delay in different Global South countries is a consistent argument on how actions taken would limit development for these same countries. More broadly behind these interlocks existed the ideology which helped develop the spread of neoliberalism. The exponential growth of neoliberal and libertarian think tanks in Global South countries from the 1990s to the early 2000s reflected the spread and expansion of expertise, knowledge exchange, and opportunities to impact policy agendas. In short, these organizations had the opportunity to become policy entrepreneurs and attempted to use this role to facilitate delayed action on climate change (Ohemeng, 2015). Thus, their campaigns represented the clear hallmarks of opposition that fed into a global interconnected CCCM.

6.4 Energy Independence, Economic Development, and South-South Cooperation

Some CCCM organizations in the USA, Europe, and others, would reposition the actions of these countries as somewhat meaningless if the rest of the world would not be limited and restricted in the same way as part of their attempt to undermine and justify delaying climate action. This position, however, neglected the historical implications of overexploitation and the unequal contributions to GHG emissions. Nevertheless, how this framing would be deployed in some organizations in Global South countries overlapped and reflected the social, political, economic, and cultural contexts of nations in the Global South.

One way in which the framing from countermovement organizations emerged is reflected in the political positions of some countries following international climate events. Following the COP15 event in 2009, before and after which India as well as other countries in the Global South such as China's positioned climate change legislation under a Polluter Pays principle (Feng & Buhi, 2010). As Jaeger and Michaelowa (2015) identified, India was a highly influential actor in international climate negotiations. Before the COP15 event, India's then prime minister implemented the Council on Climate Change and eight national missions to address energy security, before COP15 offering to reduce its emissions intensity (Thaker & Leiserowitz, 2014). Yet while undoubtedly vulnerable to the impacts of climate change, at the same time, coal and the coal industry were intrinsically linked to India's political economy (Montrone et al., 2021). This meant that the extractive industry would continue to be promoted as central to the country's development.

Arguably, this was a constructive argument reflecting the historical harms of colonial history and industrialization of the Global North. Nevertheless, how countermovement opposition then played out in these countries should still not be neglected and unchallenged. For instance, joining the voices of policy makers and government officials attempting to delay the implementation of significant GHG emissions reductions, would be the Center for Civil Society (CCS), a think tank in India and member of the Atlas Network (Chapter 3). The organization would promote what they call Terracotta Environmentalism that claimed "we approach the environment through the Terracotta vision which focuses on realigning incentives rather than reforming human nature. We believe that human action can be calibrated to positive environmental outcomes through communities, markets, and incentives" (CCS, n.d., n.p.). This is unsurprising given the organization promoted and followed the work of economists such as Hayek (e.g., CCS, 2007). While the organization did not deny climate change, its proposed responses followed from the same free market environmentalism of other neoliberal and libertarian think tanks that emerged in Global South countries. Importantly, this argument is not without its critiques and adopted by countermovement organizations in other countries affiliated with the Atlas Network. This then shows an interconnecting practice and diffusion of similar positioning and framing on the issue of climate change from organizations in the Global North to some in the Global South.

Similarly, Ordonez et al. (2021) concluded that coal was central to the Indonesian economy and energy policy. A select few companies dominated the sector; *Indonesia, ExxonMobil, Pertamina EP, Pertamina Hulu Mahakam, and the China National Offshore Oil Corporation (CNOOC)* (Migas, 2018). These were transnational fossil fuel organizations operating in key resource sectors. Thus, these outside parties in both countries and governments directly played some role in influencing climate-related policy decisions. While coal extraction may well be an important feature in the Indonesian economy, countermovement opposition also targeted the Palm oil industry. Indonesia was one of the largest palm oil producers in the world. This extractive resource was prescribed as central to development for the country (UNDP, 2019) but was entangled with land use changes that would significantly contribute to rising GHG emissions (Wicke et al., 2011).

South-South cooperation projects would facilitate the expansion and development of these industries, particularly funding from Pakistan and China. Notably, export markets between China and Pakistan were among the growing changes to trading markets connected to the Belt Road One Initiative (see Du & Zhang, 2018 for further information on Belt Road One Initiative). In addition, environmental groups were challenging the expansion of the sector which caused significant social conflict, deforestation, and consequences for climate change (Bruntse-Dahl, 2011). As such, these conflicts between environmental movements and government policy with corporate intervention (domestic and international) created the conditions required for countermovement to emerge. In effect, those wishing to protect these industries and advance development through this path conflicted with the concerns of environmental justice movements.

How this countermovement activity then manifested can be evidenced by the work of the Gecko Project (2020). A Gecko project report highlighted how like fossil fuel companies observed in the US, the palm oil industry successfully altered climate action progress. As the project noted, Indonesia made significant commitments on an international stage to address the climate crisis by setting out key reductions in deforestation. However, the actions taken following COP15 and before the Paris

Agreement in 2015 indicated a very different picture. Government officials, multinational corporations, and the financial sector all played a role in violating these approaches with the help of consultants. While less is known about the role consultants and individuals played the Project summarized that these individuals would engage in lobbying activities and could influence potential legislation, speaking on behalf of corporations. These consultants acted much like the economic consultants examined by Franta (2022) in the USA. Although, at the time of writing this book, little knowledge and detail remained on their role and if and how they utilized arguments associated with traditional CCCM organizations and actors.

Pakistan remains one of the most vulnerable countries to climate change. It was clear from early assessments that agro-ecological zones were significantly vulnerable, putting at significant risk the population. Moreover, there were low levels of adaptive capacity without international financial and economic support at a bare minimum (e.g., Malik et al., 2012). However, these clear vulnerabilities did not stop the ideological aims of one countermovement organization from operationalizing a campaign to undermine climate-related activity. The think tank Alternate Solutions Pakistan, part of the CSCCC, also sponsored some of the Heartland Institute's international climate change conferences (Heartland Institute, 2012). In fact, the organization would repost some of the same information from other think tanks or individuals from countermovement organizations in the US. For instance, reporting on Nonoy Oplas (see above) commentary they aligned his positions with that of Pakistan, exploring the importance of not restricting extractive models, particularly for expanding and allowing developing countries to export their goods; "Cutting exports by poor countries will depress their national income, undermining their capacity to import products from rich countries" (Oplas, 2008, n.p.).

Additionally, the organization would promote the Oregon Petition and the Global Warming Petition Project. The Petition Project was created in 1998 by the Oregon Institute of Science and Medicine and the George C Marshall Institute (see Chapter 2). It claimed "to demonstrate that the claim of 'settled science' and an overwhelming 'consensus' in favor of the hypothesis of human-caused global warming and consequent climatological damage is wrong" (Petition Project, 1998, n.p.) It was created by Frederick Seitz (Chapter 2) and response to the US signature to the

Kyoto Protocol. The Petition claimed to be signed by over 30,000 individuals with qualifications across scientific disciplines (Petition Project, n.d.). Alternate Solutions tied this US-specific campaign in response to the 2009 Copenhagen agreement (Athzori, 2010) but reorientated the framing to prioritize Pakistan's national interests. Yet behind these positions were the ideological aims of the organization. For example, the additional promotion by the organization of articles by Indur Goklany that critiqued activists and challenged collective action in response to the climate crisis (Goklany, 2009). As such, it became clear that delaying discourses and using contrarian's work was a tactic to fundamentally shape a narrative that collective action, state intervention, and structural economic changes would be a problem.

6.5 Industry Associations, Domestic and Transnational Organizations

Lastly, a similar pattern of industry associations and domestic and transnational corporations would set the foundations for countermovement opposition in some Global South countries. Throughout the book, the fossil industry has been described as the primary industry that initiated the operations of the CCCM in the USA, Australia, Canada, and across Europe. Importantly, fossil industry actors operate across countries; where they require access to raw materials that stimulate global commodity chains, production, and consumption of goods worldwide. Notably, the automobile sector and industry associations engaged in policy discourse to protect this industry's interests and potentially influence climate-related policy (e.g., Byrne et al., 2007; Laville, 2019; Levy & Egan, 1998; Influence Maps, n.d.) and some of these actors emerged in countries in the Global South.

Aforementioned, Brazilian think tanks that disseminated discourses of delay emerged to respond to the proposed changes in legislation to address the climate crisis, particularly in regard to agricultural reform. Moreover, these opportunities rose in tandem with changing dynamics of the Brazilian political system. Before the 1990s, the Brazilian political system could best be understood as a collation of politicians with vague and fluid ideological positions where self-interested politicians flirted between parties for their own gain (Viola & Franchini, 2014). This meant a fragmentation of Brazilian political parties with a subsequent disorganized political response and public engagement. However, from the 1990s onwards a more party-oriented legislators and a more party-pragmatic political system aligning with shifts in market reforms emerged (Borges, 2007). These changes resulted in growing political partisanship, creating opportunities for partisan divisions. In turn, elite and powerful characters could operate and integrate political narratives around land ownership, environmental policymaking and development pathways. This background presented an opportunity for similar countermovement activities observed in other countries to manifest.

In response to land ownership disputes, which became fundamental to discussions on the Forest Code and deforestation in Brazil, social movements and political parties operated on opposing sides. Like other environmental movements that pursued action to address climate change in other countries, Brazil's Movement of Landless Workers from the 1980s appeared to join the campaign. The movement promoted land distribution to poor families and aligned with left politics and revolutionary movements that reacted to the consequences of the neoliberal project (Veltmeyer & Petras, 2002). As a competing force, the *União Democrática Ruralista* (in English the Rural Democratic Union [UDR]) formed in 1985. It was made up of landowners and "defenders" of their lands that were considered endangered (Petry, 2012) by restrictions embedded in the Forest Code. For a summary of the Forest Code see Vieira et al. (2018). The UDR emerged as an expression of rural employers. Employers claimed policies were out of touch with their well-being and the UDR escalated this position in the political sphere. The organization had a history of opposition to land reforms and expropriations across Brazil. Behind its position are ideological underpinnings—not unlike those promoted by USA CCCM think tanks—of individual interests, freedom, and private property rights. Locatel and Sulva de Lima (2016) noted that the subsequent organization and development of the Brazilian political and constitutional infrastructure presented the remains of an agro-mercantile capital with colonial roots. Therefore, the Ruralistas were able to penetrate the political realm and, hypothetically, could have influenced policy protecting the agro-industry and mitigating the impact of climate-related policy. This then presented the markers of successful countermovement activity.

Furthermore, unlike the USA, corporate campaign donations are illegal in Brazil. However, many wealthy *Ruralistas* could self-fund their own political campaigns and succeed in elections (Viscidi & Graham, 2019). With this concentrated wealth and power, representatives could obtain and remain in positions that inform and influence policy decisions in Congress. For instance, between 2010 and 2015, Ruralista bench representatives dominated the conversation. They become part of key political parties and drive lobbying within the Brazilian Congress to support the agribusiness sector's expansion. This ultimately had policy implications. In 2014, Dilma Rousseff (then president) was closely aligned with the agribusiness industry (Lorris, 2018), overseeing an alliance with the Ruralista's and the pro-growth agri-sector. Dilma at the time could be considered on the left-hand side of the political spectrum as leader of the Democratic Labor Party (PDT), yet she faced significant conflict from environmental movements in response to the close relationship formed between the administration and the agribusiness sector (Vidal & Carrington, 2012). Because the agribusiness sector presented a valuable resource for increasing and expanding development in the country representing an alliance and entanglement between GDP growth, agricultural industry and developmental policy, it followed suit that policy reforms that would attempt to curb some of the ecological consequences of this industry would be challenged and stalled.

The UDR and related trade associations undertook an effective campaign to undermine climate-related legislation. For example, Blaustein and Santiago (2010) reported that, Ruralista representatives and supporters subverted the USA Avoided Deforestation Partners (ADP) narrative, which aimed to introduce carbon payments reducing greenhouse gas emissions from deforestation and forest degradation. They transformed a pro-conservation project, to the opposite in Brazil, shifting to a position that would be "costly" to Brazilian farmers. This combination of actors and the complex use of PR and lobbying was used to draw an argument of "unfair international" actors pushing policy conservation measures that would undermine economic growth and prosperity. Moreover, in 2015, the UDR helped craft the 2015 amendments to the Forest Code to reframe these within the sustainable development discourse in Brazil. It culminated in predatory agricultural practices where sustainable development goals were framed with pseudo-arguments by presenting that the environment's welfare could only be achieved through increased economic development, prosperity and reducing social and economic inequalities within the population. This was operational in the context where political forces aligned nationalistic tendencies with economic disparities and development, coalescing around the issue of deforestation that accelerated from the late 2000s. A disgruntled public and political polarization led to cuts in environmental, social, and science related governmental activities that worked via messaging on saving the interests of Brazilian farmers to help save the nation. However, these changes in environmental policy were not there to invest and save the population, but rather to preserve harmful ecological practices to protect elite interests. This messaging strategy ultimately draws back to common discourses of delay operationalized by countermovement organizations across countries.

An industry association operational in the CCCM in the USA was the International Organization for Motor Vehicle Manufacturers (OICA). The OICA board of directors included those associated with General Motors, Ford, and other large automobile manufacturers, both of which understood in the 1960s that there was some negative relationship between their industries and climate change (Beals, 2020). In addition, the OICA had affiliated member organizations worldwide (OICA, n.d.). For example, its affiliated member organization in Brazil the National Association of Automotive Vehicle Manufacturers (ANFAVEA), was founded on May 15, 1956, representing companies engaged in vehicle manufacturers. Benítez (2018) noted that ANFAVEA was a clear player in the policy networks relating to trades and tariffs that would ultimately impact climate and environmental policies.

Similarly, in Saudi Arabia, the Council of Saudi Chambers was a business association representing the fossil fuel sector and member of the OICA. It influenced domestic and outside governments through lobbying (Krane, 2020). The oil industry ultimately underpinned the country's economy and stature in international relations. Therefore, at domestic level, policies would center on expanding extractive industries and export, while at the same time in international negotiations making their position clear by vigorously pronouncing their national interests. From the late 1990s to the early 2000s the Saudi government, and I add accompanying lobbying groups, would aim to undermine the science and deploy narratives that would feed into scientific doubt as well as centering the negative economic impacts of diversifying the energy supply and reducing extractive industries (Depledge, 2008). This is despite the country's vulnerability to the impacts of climate change.

6.6 Conclusion

Countermovements respond to prevailing social, cultural, economic, and political challenges. While political economies may have been distinct and different from that of other countries mentioned in this book, in some countries in the Global South we see evidence of similar organizations and individuals, particularly from the 2000s connected to countermovement organizations in the USA and Europe. Moreover, organizations operating in similar roles to trade associations in the US operationalized activities to undermine policy developments to reduce GHG emissions. This initial insight presents only a glimpse into the history of countermovement opposition. Nevertheless, there are four key conclusions that we can draw.

One, there was a notable network of libertarian and neoliberal think tanks that emerged as part of a policy project to expand and influence politics with a particular ideological aim. While in different social, political, and economic contexts that would ultimately affect the narratives of each organization, what was consistent was these interconnections. Two, individuals cited by or engaged by different organizations, would spread similar and consistent discourses of delay acting as legitimate and authoritative voices much like other countries identified in this book. Three, solutions to the realities and vulnerabilities of countries to climate change centered in some cases of ecological modernization, facilitating increased growth and solutions to protecting the environment. Four, this focus on development would intersect with different trading partnerships particularly those of South-South cooperation expanding export markets to decrease poverty and become a liberating movement. There is a balancing argument to take that this development argument is constructive. Yet, it remains that there are clear overlaps to CCCM organizations in other countries identified in this book where similar or amended opposition activities and tactics would be deployed to delay policy efforts to mitigate climate change. Thus, this chapter has only scratched the surface of what we know about the countermovement in the Global South and at the time of writing, scholarship and understanding is still ongoing.

References

Association of Liberal Thinking. (n.d.). *Center for Environmental Studies*. http://www.liberal.org.tr/page/kat.php?kat=6&akat=15. Accessed 1 Oct 2022.

Athzori, D. (2010). *Copenhagen: The skeptics view*. https://asinstitute.org/copenhagen-the-skeptics-views/. Accessed 1 Oct 2022.

Ayodele, T. (2009). *North Hides Nefarious aims under Green Cloak*. http://www.ippanigeria.org/articles/North%20Hides%20Nefarious%20Aims%20Under%20Green%20Cloak.html. Accessed 1 Oct 2022.

Beals, R. K. (2020, October 31). *GM and Ford knew about climate change 50 years ago, energy trade publication has uncovered*. https://www.marketwatch.com/story/gm-and-ford-knew-about-climate-change-50-years-ago-energy-trade-publication-has-uncovered-11603738820. Accessed 1 Oct 2022.

Benítez, G. J. (2018). Business lobbying: Mapping policy networks in Brazil in Mercosur. *Social Sciences, 7*(10), 198. https://doi.org/10.3390/socsci7100198

Blaustein, R., & Santiago, C. (2010). *Will Brazil change its forest code and kills the Amazon*. https://www.ecosystemmarketplace.com/articles/will-brazil-change-forest-code-kill-amazon/. Accessed 10 Dec 2022.

Bogert, J., Ellers, J., Lewandowsky, S., Balgopal, M., & Harvey, J. (2022). Reviewing the relationship between neoliberal societies and nature: Implications of the industrialized dominant social paradigm for a sustainable future. *Ecology and Society, 27*(2), 7.

Bordner, A. S., Ferguson, C. E., & Ortolano, L. (2020). Colonial dynamics limit climate adaptation in Oceania: Perspectives from the Marshall Islands. *Global Environmental Change, 61*, 102054.

Borges, A. (2007). Rethinking state politics: The withering of state dominant machines in Brazil. *Brazilian Political Science Review, 1*(2), 108–136.

Bruntse-Dahl, R. (2011, May 20). Malaysia and Indonesia bolster defence of palm oil industry to west. *The Guardian*. https://www.theguardian.com/environment/2011/may/20/malaysia-indonesia-palm-oil-defence. Accessed 10 Dec 2022.

Byrne, J., Hughes, K., Rickerson, W., & Kurdgelashvili, L. (2007). American policy conflict in the greenhouse: Divergent trends in federal, regional, state, and local green energy and climate change policy. *Energy Policy, 35*(9), 4555–4573.

Cato Institute. (n.d.). *Remembering Giancarlo Ibarguen 1963–2016*. https://www.cato.org/blog/remembering-giancarlo-ibarguen-1963-2016. Accessed 1 Oct 2022.

Center for Civil Society. (2007). *About us*. https://web.archive.org/web/20070222065306/http://www.ccs.in/aboutus.asp. Accessed 1 Oct 2022.

Center for Civil Society. (n.d.). *Terracotta Environmentalism*. https://ccs.in/node/485. Accessed 1 Oct 2022.

Champéroux, A. (2019). The birth and foundation of an international 'polluter pays' principle. *Internetowy Przegląd Prawniczy TBSP UJ, 4*(48), 33–45.
Coelho-Junior, M. G., Valdiones, A. P., Shimbo, J. Z., Silgueiro, V., Rosa, M., Marques, C. D. L., Oliveira, M., Araújo, S., & Azevedo, T. (2022). Unmasking the impunity of illegal deforestation in the Brazilian Amazon: A call for enforcement and accountability. *Environmental Research Letters, 17*(4), 041001.
Conservative Transparency. (n.d.). *Pierre F. and Enid Goodrich Foundation*. http://conservativetransparency.org/donor/pierre-f-and-enid-goodrich-foundation/. Accessed 29 Nov 2022.
Datta, A., & Young, J. (2011). Producing home grown solutions: Think tanks and knowledge networks in international development. *Development Outreach, 13*(2), 34–37.
Davies, K. (2016). Competitive Enterprise Institute NYT Ad Signatories Got $10 Million from Exxon. https://climateinvestigations.org/nyt_ad_signatories_got_10_million_from_exxon/. Accessed 1 Oct 2022.
De Smog. (n.d.). *Instituto Liberdade*. https://www.desmog.com/instituto-liberdade/. Accessed 29 Nov 2022.
Depledge, J. (2008). Striving for no: Saudi Arabia in the climate change regime. *Global Environmental Politics, 8*(4), 9–35.
Doyle, T., & Lockhart, A. (2012). Local campaigns against Shell or transnational campaigns against climate change? From the Niger Delta, Nigeria to Rossport, Ireland. In *Local climate change and society* (pp. 182–203). Routledge.
Du, J., & Zhang, Y. (2018). Does one belt one road initiative promote Chinese overseas direct investment? *China Economic Review, 47*, 189–205.
Eberlein, R. (2006). On the road to the state's perdition? Authority and sovereignty in the Niger Delta, Nigeria. *The Journal of Modern African Studies, 44*(4), 573–596.
Economic Freedom of the World. (n.d.). *List of organizations*. https://www.fraserinstitute.org/economic-freedom/our-network/member-institutes. Accessed 1 Oct 2022.
European Parliament. (2022). *Brazil*. https://www.europarl.europa.eu/RegData/etudes/BRIE/2022/738185/EPRS_BRI(2022)738185_EN.pdf. Accessed 10 Dec 2022.
Feng, L., & Buhi, J. (2010). The Copenhagen Accord and the silent incorporation of the Polluter Pays Principle in International Climate Law: An analysis of Sino-American diplomacy at Copenhagen and Beyond. *Buffalo Environmental Law Journal, 18*, 1.
Foster, J. B., & Clark, B. (2004). Ecological imperialism: The curse of capitalism. *Socialist Register*, *40*.

Foster, J. B., & Holleman, H. (2014). The theory of unequal ecological exchange: A Marx-Odum dialectic. *Journal of Peasant Studies, 41*(2), 199–233.

Fotopoulos, T. A. K. I. S. (2008). Values, the dominant social paradigm and neoliberal globalization. *The International Journal of Inclusive Democracy, 4*(1), 1–18.

Franta, B. (2022). Weaponizing economics: Big Oil, economic consultants, and climate policy delay. *Environmental Politics, 31*(4), 555–575.

Givens, J. E., Huang, X., & Jorgenson, A. K. (2019). Ecologically unequal exchange: A theory of global environmental injustice. *Sociology Compass, 13*(5), e12693.

Goklany, I. (2008). *What to do about climate change*. https://www.elcato.org/que-hacer-con-el-cambio-climatico. Accessed 1 Oct 2022.

Goklany, I. (2009). *Apocalypse when*. https://asinstitute.org/apocalypse-when/. Accessed 1 Oct 2022.

Gore, C. (2000). The rise and fall of the Washington Consensus as a paradigm for developing countries. *World Development, 28*(5), 789–804.

Gray, K., & Gills, B. K. (2016). South-South cooperation and the rise of the Global South. *Third World Quarterly, 37*(4), 557–574.

Heartland Institute. (2012). *Cosponsors*. https://web.archive.org/web/20120623162748/http://climateconference.heartland.org/cosponsors/. Accessed 1 Oct 2022.

Influence Maps. (n.d.). *The automobile sector and climate change*. https://influencemap.org/report/The-Automotive-Sector-and-Climate-Change-18218. Accessed 1 Oct 2022.

Initiative for Public Policy Analysis. (2004). *Projects and activities*. https://web.archive.org/web/20040803205604/http://ippanigeria.org/projects_and_activities.php. Accessed 1 Oct 2022.

Instituto Libderdade. (2010). https://web.archive.org/web/20101201094916/http://www.il-rs.org.br/ingles/info/climate.php. Accessed 1 Oct 2022.

Instituto Liberdade. (n.d.). *About*. https://institutoliberdade.com.br/. Accessed 1 Oct 2022.

International Organization for Motor Vehicle Manufacturers. (n.d.). Members. https://www.oica.net/category/about-us/members/. Accessed 1 Oct 2022.

Jaeger, M. D., & Michaelowa, K. (2015). Energy poverty and policy coherence in India: Norms as means in a strategic two-level discourse. In M. Knodt, N. Piefer, & F. Müller (Eds.), *Challenges of European External Energy Governance*. Ashgate.

Jorgenson, A. K. (2006). Unequal ecological exchange and environmental degradation: A theoretical proposition and cross-national study of deforestation, 1990–2000. *Rural Sociology, 71*(4), 685–712.

Kalu, K. A. (2006). The international dimension of regime change: Economic sanctions and the 1993 democratic transition in Nigeria. *African Social Science Review, 3*(3), 1.

Kolk, A., & Levy, D. (2001). Winds of change: Corporate strategy, climate change and oil multinationals. *European Management Journal, 19*(5), 501–509.

Krane, J. (2020). Climate strategy for producer countries: The case of Saudi Arabia. *When can oil economies be deemed sustainable?* 301.

Lau, L. C., Tan, K. T., Lee, K. T., & Mohamed, A. R. (2009). A comparative study on the energy policies in Japan and Malaysia in fulfilling their nations' obligations towards the Kyoto Protocol. *Energy Policy, 37*(11), 4771–4778.

Laville, S. (2019, October 10). Exclusive: Carmakers among key opponents of climate action. *The Guardian*. https://www.theguardian.com/environment/2019/oct/10/exclusive-carmakers-opponents-climate-action-us-europe-emissions. Accessed 1 Oct 2022.

Levy, D. L., & Egan, D. (1998). Capital contests: National and transnational channels of corporate influence on the climate change negotiations. *Politics & Society, 26*(3), 337–361.

Liberty Fund. (n.d.). *Malthusianism, resourceship and Liberty Conference*. https://www.libertyfund.org/conference/malthusianism-resourceship-and-liberty/. Accessed 1 Oct 2022.

Liggio, L. P. (1990, January 12). The Hispanic tradition of liberty. The road not taken in Latin America. In *Mont Pelerin Society Regional Meeting*. https://leonardliggio.org/wp-content/uploads/downloads/2011/10/THE%20HISPANIC%20TRADITION%20OF%20LIBERTY-%20THE%20ROAD%20NOT%20TAKEN%20IN%20LATIN%20AMERICA%201-12-1990.pdf. Accessed 1 Dec 2022.

Lingle, C. (2003). *Global warming: Extreme weather or extreme prejudice?* https://fee.org/articles/global-warming-extreme-weather-or-extreme-prejudice/. Accessed 1 Oct 2022.

Lingle, C. (2004, September 1, Wednesday). Higher CO_2, more global warming, and less extinction? *Foundation for Economic Education*. https://fee.org/articles/higher-co2-more-global-warming-and-less-extinction/. Accessed 1 Oct 2022.

Locatel, C. D., & de Lima, F. L. S. (2016). Territórios rurais e reestruturação produtiva do capital no Rio Grande do Norte. *Formação (Online), 1*(23).

Loris, A. A. (2018). Seeding a narrow future and harvesting an exclusionary past: The contradictions and future scenarios of agro-neoliberalism in Brazil. *Futures, 95*, 76–85.

Malik, S. M., Awan, H., & Khan, N. (2012). Mapping vulnerability to climate change and its repercussions on human health in Pakistan. *Globalization and Health, 8*(1), 1–10.

Manby, B. (1999). The role and responsibility of oil multinationals in Nigeria. *Journal of International Affairs, 53*(1), 281–301.

McKie, R. E. (2021). Obstruction, delay, and transnationalism: Examining the online climate change counter-movement. *Energy Research & Social Science, 80*, 102217.

Michaels, P (2010). *Inconvenient truths of Climategate*. https://web.archive.org/web/20101128032515/http://www.elcato.org/node/4797. Accessed 1 Oct 2022.

Migas, S. K. K. (2018, October 11). What are the biggest oil & gas companies in Indonesia? *Indonesia Investments*. https://www.indonesia-investments.com/news/todays-headlines/what-are-the-biggest-oil-gas-companies-in-indonesia/item9000? Accessed 1 Oct 2022.

Milani, C. R., & Chaves, L. N. (2022). How and why European and Chinese pro-climate leadership may be challenged by their strategic economic interests in Brazil. *Asia Europe Journal, 20*(4), 403–422.

Mirowski, P. (2014). *The political movement that dared not speak its own name: The neoliberal thought collective under erasure*. Institute for New Economic Thinking Working Paper Series No. 23.

Mirowski, P., & Plehwe, D. (Eds.). (2015). *The road from Mont Pèlerin: The making of the neoliberal thought collective, with a new preface*. Harvard University Press.

Mont Perelin Society. (2007). *Home page*. https://web.archive.org/web/20070809194715/http://www.montpelerin.org/home.cfm. Accessed 1 Oct 2022.

Montrone, L., Ohlendorf, N., & Chandra, R. (2021). The political economy of coal in India-Evidence from expert interviews. *Energy for Sustainable Development, 61*, 230–240.

Nassau Institute. (2002). *Global warming—Cool Off on the Global Warming Rhetoric*. https://www.nassauinstitute.org/article232/. Accessed 9 Dec 2022.

Nassau Institute. (2003). Global Warming: What should we be doing about it? https://www.nassauinstitute.org/article8/. Accessed 28 Oct 2022.

Nassau Institute. (2004). *Global warming—What should we be doing about it*. https://web.archive.org/web/20040831014726/http://www.nassauinstitute.org/wmview.php?ArtID=312. Accessed 9 Dec 2022.

Nassau Institute. (2009). *Is The Nassau Institute dead wrong about climate change?* https://www.nassauinstitute.org/article854/. Accessed 9 Dec 2022.

Nassau Institute. (n.d.). *Homepage*. https://www.nassauinstitute.org/. Accessed 9 Dec 2022.

Ohemeng, F. L. K. (2015). Civil society and policy making in developing countries: Assessing the impact of think tanks on policy outcomes in Ghana. *Journal of Asian and African studies, 50*(6), 667–682.

Omeje, K. (2017). Extractive economies and conflicts in the global south: Re-engaging rentier theory and politics. In *Extractive economies and conflicts in the Global South* (pp. 19–44). Routledge.

Oplas, N. (2008). Make the poor pay. https://asinstitute.org/make-the-poor-pay-4/. Accessed 22 Oct 2022.

Oplas, N. (2009). *The Sun vs CO_2: What drives global temperature changes?* https://web.archive.org/web/20091025033910/http://www.minimalgovernment.net/media/mg_20090327.pdf. Accessed 9 Dec 2022.

Oplas, N. (2010). *Essays on climate alarmism*. https://web.archive.org/web/20110915214203/http://www.minimalgovernment.net/media/mg_20100630.pdf. Accessed 10 Dec 2022.

Ordonez, J. A., Jakob, M., Steckel, J. C., & Fünfgeld, A. (2021). Coal, power and coal-powered politics in Indonesia. *Environmental Science & Policy, 123*, 44–57.

Petition Project. (1998). *Homepage*. http://www.petitionproject.org/index.php. Accessed 9 Dec 2022.

Petition Project. (n.d.). *Qualifications of signatories*. http://www.petitionproject.org/qualifications_of_signers.php. Accessed 9 Dec 2022.

Petry, C. A. (2012). Atuação da bancada ruralista nas votações de projetos relacionados ao Novo Código Florestal Brasileiro durante o governo Dilma. http://hdl.handle.net/10183/67249. Accessed 9 Dec 2022.

Plehwe, D. (2014). Think tank networks and the knowledge–interest nexus: The case of climate change. *Critical Policy Studies, 8*(1), 101–115.

Plehwe, D. (2015). Introduction. In P. Mirowski & D. Plehwe (Eds.), *The road from Mont Pèlerin: The making of the neoliberal thought collective, with a new preface*. Harvard University Press.

Polonsky, M., Kilbourne, W., & Vocino, A. (2014). Relationship between the dominant social paradigm, materialism and environmental behaviors in four Asian economies. *European Journal of Marketing, 48*(3/4), 522–551.

Rajão, R., Soares-Filho, B., Nunes, F., Börner, J., Machado, L., Assis, D., ... & Figueira, D. (2020). The rotten apples of Brazil's agribusiness. *Science, 369*(6501), 246–248.

Saputra, M. D. H. (2019). Beyond Neoliberalism, the emergence of Latin America Pink Tide (Post Washington Consensus and Neostructuralism). *Review of International Relations, 1*(1). https://doi.org/10.24252/rir.v1i1.9305

Thaker, J., & Leiserowitz, A. (2014). Shifting discourses of climate change in India. *Climatic Change, 123*(2), 107–119.

The Gecko Project. (2020). The consultant: Why did a palm oil conglomerate pay $22m to an unnamed 'expert' in Papua? https://news.mongabay.com/2020/06/the-consultant-why-did-a-palm-oil-conglomerate-pay-22m-to-an-unnamed-expert-in-papua/. Accessed 1 Oct 2022.

UN. (n.d.). *What is 'South-South cooperation' and why does it matter*. Department of Economic and Social Affairs. https://www.un.org/development/desa/en/news/intergovernmental-coordination/south-south-cooperation-2019.html. Accessed 1 Oct 2022.

UNDP. (2019). *Indonesia: At a glance country guide*. https://www.undp.org/facs/publications/indonesia-glance-country-guide. Accessed 1 Oct 2022.

UNFCCC. (n.d.-a). *Bahamas*. https://unfccc.int/node/61009. Accessed 1 Oct 2022.

UNFCCC. (n.d.-b). *Parties and observers*. https://unfccc.int/parties-observers. Accessed 1 Oct 2022.

Veltmeyer, H., & Petras, J. (2002). The social dynamics of Brazil's rural landless workers' movement: Ten hypotheses on successful leadership. *Canadian Review of Sociology/Revue anadienne de sociologie, 39*(1), 79–96.

Vidal, J., & Carrington, D. (2012, May 11). Petition calls on Brazilian president to veto catastropgic forest code. *The Guardian*. https://www.theguardian.com/environment/2012/may/11/petition-brazil-president-veto-forest-code

Vieira, R. R. S., Ribeiro, B. R., Resende, F. M., Brum, F. T., Machado, N., Sales, L. P., & Loyola, R. (2018). Compliance to Brazil's Forest Code will not protect biodiversity and ecosystem services. *Diversity and Distributions, 24*(4), 434–438.

Viola, E., & Franchini, M. (2014). Brazilian climate politics 2005–2012: Ambivalence and paradox. *Wiley Interdisciplinary Reviews: Climate Change, 5*(5), 677–688.

Viscidi, L., & Graham, N. (2019, January 4). Brazil was a global leader on climate change. Now it's a threat. *Foreign Policy*.

Walker, J. (2020). Neoliberalism, environmentalism, and the crisis of the 1970s. *More heat than life: The tangled roots of ecology, energy, and economics* (pp. 3–31). Palgrave Macmillan.

Wicke, B., Sikkema, R., Dornburg, V., & Faaij, A. (2011). Exploring land use changes and the role of palm oil production in Indonesia and Malaysia. *Land Use Policy, 28*(1), 193–206.

Williamson, J. (1990). *The Washington consensus*.

CHAPTER 7

Conclusion

When writing this book, the COP26 event occurred in Glasgow, United Kingdom. Thirty years after the first UNFCCC conference in Rio de Janeiro, it became clear that the historical implications of obstructive efforts to undermine climate policy had stalled international agreements, undermined nationally determined contributions and GHG emissions targets, and as the IPCC ARC 6th Report revealed that the world would probably exceed the initial 1.5 degree increase in Earth's Temperature (World Resource Institute, 2021). How much of this failure at international negotiations to implement national-level policy can be placed on the role of countermovement opposition is unknown. Nevertheless, this book has described that over thirty years, the activities of those invested in preventing action on climate change worked extremely hard to influence policy decisions and public opinion.

Evidence throughout this text presents the significant role of the CCCM that has actively sought, over the past 30 years, to undermine progressive action on climate change mainly on behalf of the fossil fuel industry. The USA was the birthplace of this opposition movement. Using money to build a network of neoliberal and libertarian think tanks to disseminate their messages and making PAC contributions to politicians and political parties, the fossil fuel industry aimed to influence the political process. Together with coalitions, trade associations, and an accompanying media echo chamber ensured a stream of delaying discourses would

R. E. McKie, *The Climate Change Counter Movement*,
https://doi.org/10.1007/978-3-031-33592-1_7

penetrate politics and the public minds. Put the two together and US opposition were certainly able to delay progressive climate action in the USA.

Yet this does not necessarily explain how outside of the USA, similar and connected organizations and individuals would undertake countermovement activities to undermine climate change action in other countries. Climate change is a global issue, while not as well organized as it was in the USA, organizations would build a platform of countermovement opposition to present a consistent yet diverse range of arguments to intercept political discourse and public opinion. What connects the US opposition to much of the other countries is the ideological foundations of countermovement opposition. In short, ideological the positions of CCCM organizations are underpinned by, on varying levels, to support deregulation, limited to no government intervention, and a more rigid expansion of marketization of all elements of the economy. This is then reflected in their opposition to increased state regulation associated with reducing GHG emissions. This is accompanied by a resistance to collectivism articulated by environmental and climate justice movements.

This ideological influence is clear in the case of European countermovement opposition. Emerging across Europe was the strategic alliance of think tanks and individuals that operationalized discourses of delay across and beyond Europe to spread ideas and doubt to impact legislation and public opinion; some of these think tanks were directly related to the fossil fuel industry. The alliance of think tanks would not have succeeded if it were not for Anthony Fisher. Fisher would craft a network of think tanks first facilitating the rise of the IEA, the IPN before helping to create the Atlas Network which would recruit and support the rise of think tanks across every continent. This network built a cohesive policy project aimed at reducing government intervention in most policy sectors which included the extractive industries. Thus, it reflected a spread of ideas intersecting with the political and economic shifts to a neoliberal global society and the spread of ideas from politics into policy. Spreading ideas on political issues would be a central ploy of those with vested interests in protecting the status quo with many of these think tanks in Europe. This network expanded into Australia, Canada, and some countries in the Global South. These think tank networks then consolidated messages that could be diffused through countries and disseminated into the political realm.

Accompanying this spread of ideas, it was the particular role of individuals to disseminate discourses of delay or spread and penetrate different CCCM organizations across countries. Climate contrarians, or individuals such as specific economists, consultants, and others acted among this global network. From the PR campaign and lobbying activities of Fritts Botcher in Europe to the repeated use of contrarian works such as Bob Carter and Patrick Michael, contrarian's or specific individuals played a crucial role in providing authoritative voices for this opposition movement. These individuals also interlocked with industry associations representing heavily polluting industries such as Viv Forbes.

Importantly, while the initial campaign was set up and led by fossil fuel corporations, different corporate and state interests related to other heavily polluting industries would become vocal opponents of climate action. For example, in New Zealand and Brazil, organizations and individuals related to the agribusiness sector would use the same tactics to undermine climate action noting the growing emissions from these industries. The result was the growth of an international countermovement opposition. However, there were complexities between different organizations, individuals, and across countries. This serves to illustrate a pluriverse whereby different actors in different countries would adopt similar and different approaches to explaining the one singular issue of climate change.

While there were some consistencies in the operations of CCCM organizations, further complexities continue to be a question for academic researchers. This includes finding out more about if and what these opposition movements looked like in the Global South, what the emerging changes in factors influencing opposition movements, and what new challenges would present for the countermovement that may have to adapt to the ever-growing scientific and physical evidence of damage caused by human's impact on the environment.

Following the COVID-19 pandemic, it served as a period of time to reflect on the fragility of the human population to environmental changes as well as modes of behavior that would include the reduction of personal consumption such as individual car use. Later, following the Russian invasion of Ukraine, attention turned to the fragility of a fossil fuel global economy and the risks posed to human life where energy demand and energy supply were threatened. At the same time, the pandemic response across countries would be complex and reflect the changing political dynamics of countries. The responses clearly reflected a political economic shift within countries in the Global North, where growing nationalistic tendencies would ultimately stir up political differences and social discontent. Importantly, the images associated with the DSP of material abundance and overconsumption became increasingly challenged.

All was not lost in progress made on actions to address the climate crisis. More specifically, we have observed accelerating and globalizing climate justice movements worldwide that seek to draw attention to the issue of climate change. These campaigns offer hope that despite the organized campaigns of corporate and state interests that sought to prevent action on climate change would no longer go unchallenged. Importantly, intergenerational climate justice movements seeking to bring together young people who will inhabit the planet as these climate changes become even more extreme, and older generations who recognize the need for growing action, signals that successful collective mobilization that drives social change is quite possible.

Yet individual behavior can go so far, and these collective movements must continue to challenge the multi-million dollar corporations and financial organizations that seek to protect the specific political economy that as I argue cannot be divorced from environmental degradation. Further delays could be significant without collective mobilization challenging the changing techniques of countermovement opposition. Therefore, I urge you as the reader to use this text as food for thought. What have things looked like in the past and how can you help drive change?

Reference

World Resource Institute. (2021). The IPCC climate report. https://www.wri.org/insights/ipcc-climate-report. Accessed 11 December 2022.

Index

R. E. McKie, *The Climate Change Counter Movement*,
https://doi.org/10.1007/978-3-031-33592-1

FSC
www.fsc.org
MIX
Papier | Fördert gute Waldnutzung
FSC® C083411

Kolibri 360 GmbH
Ferdinand-Jühlke-Straße 7
99095 Erfurt, Deutschland
produktsicherheit@kolibri360.de